Developing Global Marketing Strategy

Postgraduate Certificate in Global Marketing Practice

Developing Global Marketing Strategy

First edition September 2010

ISBN: 9780 7517 8411 4
eBook: 9780 7517 8422 0

British Library Cataloguing-in-Publication Data
A catalogue record for this book has been
applied for from the British Library

Published by

BPP Learning Media Ltd
Aldine House, Aldine Place
London W12 8AA

www.bpp.com/learningmedia

Printed in Great Britain

We are grateful to the GMN Advisory Council
members and Module Advisors.

Author:
Darrell Kofkin

Academic reviewer:
Dr Kellie Vincent

Your learning materials, published by
BPP Learning Media Ltd, are printed on paper
sourced from sustainable, managed forests.

A note about copyright

Contents

BPP
LEARNING MEDIA

Congratulations

on choosing to study towards MSc Global Marketing Practice.

You have made the right decision because you have at your fingertips a wealth of support to help guide you through your studies with BPP Learning Media, Ashcroft International Business School and Global Marketing Network.

GMN's CEO Darrell Kofkin summarises the key value of The Global Marketer Programme as

"The programme is designed to prepare you thoroughly for a rewarding career in marketing and ensure you can make a profitable contribution to business anywhere in the world."

For the first time, you can achieve an academic award at Masters level from one of the UK's largest and most progressive universities, while at the same time ensuring you meet all the academic criteria required for acceptance as a Professional Member of Global Marketing Network.

It is our intention that at this first level of this innovative Masters programme, Postgraduate Certificate Global Marketing Practice, you are able to develop the skills you require to perform effectively from day one.

This Study Text is designed to work in conjunction with the following elements of the overall programme:

- A Student Handbook designed to give you all the information you need to achieve success

- A personal E-Tutor to guide, encourage and develop your thinking and provide guidance for your assignment

- Access to The Global Marketer virtual learning environment providing you with a week-by-week plan of study for each module, with access to articles, case studies, additional learning activities and a global discussion forum connecting you with other programme participants

- Access to the Anglia Ruskin University digital library, providing access to an extensive range of online databases, world leading journals and E-books

- Access to the GMN online network with all the latest news about approved conferences, workshops and events that you can attend to support your continuing development

Each chapter of the Study Text is designed to refer to each Study Session within the programme.

Integrating Global Marketing Network Philosophies

Each module has at least one GMN Advisory Council or Faculty member acting as Module Advisor. Module Advisors contribute to the development of the curriculum and study materials.

The GMN Module Advisors for Developing Global Marketing Strategy are Prof. Svend Hollensen and Neil Stevenson.

Recognised as one of the world's leading authorities on global marketing strategy **Svend** commenced his career in international marketing for a large Danish multinational enterprise before receiving his PhD from Copenhagen Business School in 1992. With sales of 60,000 worldwide his book on global marketing strategy, *Global Marketing – A Decision Oriented Approach* is the bestselling text on the subject outside the USA. Svend now divides his time as an academic, an author, a consultant and around the world as a speaker on global marketing issues.

Neil is Executive Director of Brand for the worldwide body Association of Certified Chartered Accountants (ACCA). He has more than 10 years' experience in financial and professional services marketing. He has been an employee at ACCA since 2001, establishing ACCA's first integrated marketing and promotions team in the corporate headquarters.

Additional information and reviews of the Study Materials by Module Advisors can be found on the Global Marketer Programme supporting Virtual Learning Environment.

Using this Study Text

This Study Text includes features designed specifically to make learning effective and efficient.

- Each chapter begins with an **introduction** to set the chapter in context.

- Chapter **learning outcomes** summarise what you are expected to be able to achieve through reading the chapter. The content covered is summarised in a Chapter Roundup which is organised according to these learning outcomes at the end of the chapter.

- Harvard referencing is used throughout with **a full reference list** at the end of each chapter.

Throughout the Study Text, there are also special aids to learning. These are indicated by the following symbols.

Definition

Definitions are provided for key terms.

Activity 1

Activities for you to complete are included througout the chapters. Often these will ask you to relate theories to your own organisation. Where appropriate activity debriefs are included at the end of the chapter.

Global case study

Real life global examples are used to illustrate marketing practice.

 Readily available key text book resources, significant chapters and additional reading are selectively added to the chapters to help develop your further reading.

 Selected online materials which link to the topics covered within the chapter are suggested at appropriate points.

Assessment advice

Points to bear in mind in relation to specific chapter topics when preparing your assignment are inserted at relevant places in the chapter.

GMN viewpoint

GMN experts' opinions, snippets from presentations and papers are included to bring both practitioner and academic cutting edge perspectives into the material.

About this module

This module is work-based in its nature and learning approach and builds on your existing practice-based awareness of global marketing issues. The module seeks to further your understanding of these by providing a critical and, where necessary, targeted insight into contemporary global marketing theory and practice in order that they can develop, recommend, implement and monitor a global marketing strategy for an organisation.

1. Complex global issues

This module enables you to engage practically with complex global marketing issues faced by organisations. Issues to be examined include whether to internationalise; which markets to enter; alternative market entry strategies; designing the global marketing programme; and implementing and coordinating the global marketing programme.

2. Cross cultural issues

The module will also consider the cross-cultural issues companies face in globalising their operations. The development of responses to these issues will draw upon the models and theories of working across cultures including anthropology, organisational behaviour, sociology and psychology.

3. Integrating theory and practice

You will be expected to work on an assignment related either to your own organisation or a national/international business organisation. This will ensure a clear integration between theory and practice. In order to complete this assignment, you should draw upon a range of theories relating to the discipline of Global Marketing.

Module aims

Both the study mode and module content further facilitate the application of, and reflection on, theory in practice and these are evidenced in the learning outcomes and the associated development of the individual's understanding and organisational impact of the methods and approaches necessary to develop, implement and monitor an effective global marketing strategy for the organisation.

The module supports the development of relevant employability, promotability and professional skills which enables you to develop an understanding of the organisation.

Learning outcomes for the module are to:

1   Demonstrate a critical awareness of global marketing models/theories and issues caused by working with and in different countries/cultures.

2   Critically evaluate appropriate market entry strategies for organisations considering global expansion, and their impact on the organisation's existing strategy, resources, processes and systems.

3   Develop a marketing plan that supports the global marketing strategy.

4   Analyse the required performance and monitoring framework that measures and controls the global marketing strategy.

5   Assess the importance of cross-cultural issues and their impact on the successful implementation of the global marketing strategy.

6   Reflect critically upon own learning throughout the module, and articulate the value of this learning to an organisation's and to the student's own professional development.

A note on pronouns

On occasions in this Study Text, 'he' is used for 'he or she', 'him' for 'him or her' and so forth. While we try to avoid this practice it is sometimes necessary for reasons of style. No prejudice or stereotyping accounting to sex is intended or assumed.

Key contacts

Please email gmn@bpp.com for any queries about this Study Text and The Global Marketer Programme.

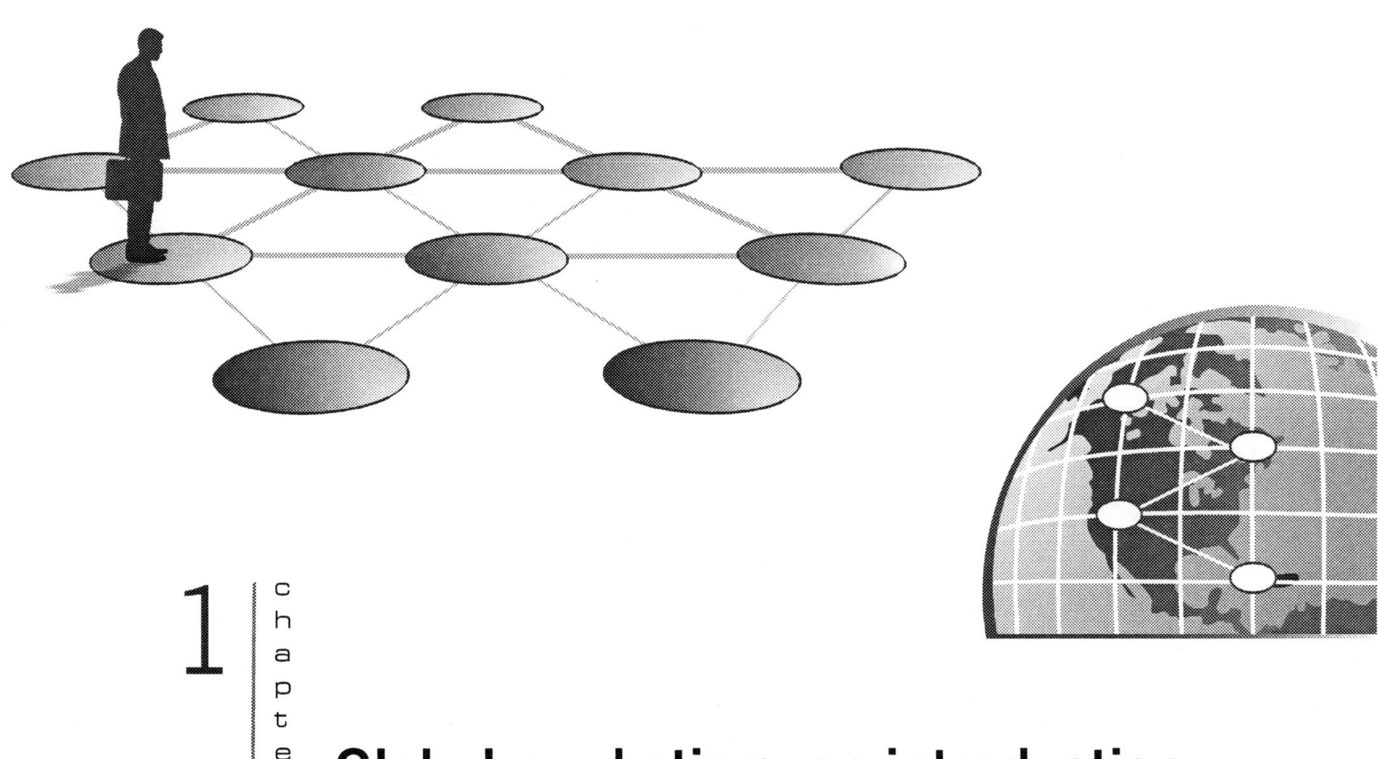

1 | chapter

Global marketing: an introduction

In the face of globalisation, and with many organisations under increasing competitive pressures, there are moves within organisations to strengthen their capabilities to enable them to move into new foreign markets.

The development and implementation from a domestic to a global organisation is unlikely to succeed without proper planning. It is a complex process which requires creativity, adaptability and an openness to the uncertainty that global expansion offers. Get it right and the organisation will achieve tremendous success. Get it slightly wrong and the results can be disastrous.

This chapter provides an introduction to the concept of global marketing strategy.

We begin with an overview of the importance of international trade, the major motivations of the organisation in embarking on a global marketing strategy, some of the central themes and concepts that explain the organisation's process of internationalisation and current issues in global marketing strategy including the emergence of the 'born global' organisation.

This chapter also looks at the impact of international trade and how it develops, presents an overview on the growing importance of emerging markets and examines the increasing importance of technology in global marketing strategy development.

Contents

Chapter learning outcomes

In this chapter you will cover the following:

- Consider the impact of international trade
- Understand motives and triggers of internationalisation
- Evaluate the development of the global marketing concept
- Understand the evolution of global marketing strategy

1 The impact of international trade

1.1 Overview

Without international trade, there would be no global marketing. The complexities of international trade mean that you can only very rarely regard the development and implementation of a global marketing strategy as an extension of a domestic marketing strategy. Indeed organisations that do this, and see it as a bolt on or a separate strategy without integrating it into their current organisation, rarely, if, ever, succeed.

At one time it was felt that countries were independent economic units and that trade did not affect the whole economy. International trade has the potential to make all parties better off for the same amount of effort. Recent international agreements have prodded the world towards freer trade but there is still some way to go to a genuinely global market in every sector.

1.2 The importance of 'going global' for organisations

Any business pursues **profits**. An overseas market or the global market offers larger profits.

- Increased **volume sales** overall
- The ability to **specialise** more
- The ability to **segment a wider market**
- The ability to **draw resources** from a wider market

For example, the UK domestic market for aircraft (both civilian and military) could not support an aerospace industry as the demand would not be big enough, but exports enable such companies to be profitable. Small and medium sized enterprises too can benefit from export markets.

Many firms could not survive at all without international trade. Volvo and Saab, the Swedish car manufacturers, sell by far the majority of their output in overseas markets, as Sweden would be too small to support them. Some of the world's largest chemicals companies are Swiss, yet demand in Switzerland is not big enough to sustain them.

All companies are affected by economic forces. **Different firms respond in a variety of ways to these forces.**

- Some **ignore overseas markets** and concentrate on their domestic market
- Some **export** to overseas markets
- Some **internationalise** their marketing and their production

This is because overseas markets are not only a place where customers can be found, but a source of resources with which customers can be satisfied. A firm based abroad, acting as an international business, has many more sources of supply than one trading only at home.

1.3 The importance of international trade for countries

The environment of international trade is controlled and policed by national governments directly or as a result of treaties. Most governments believe that international trade is beneficial.

(a)   A country normally **exports** goods which it can produce most efficiently, that is, at lowest cost relative to other countries, and it normally imports goods in which other countries are relatively more efficient. This is called the principle of **comparative advantage**. World output is larger when each country **specialises** in goods it can produce most efficiently.

(b)   The goods in which a country will have comparative advantage depend partly on supplies of the **factors of production**. A country with much richer agricultural land will tend to export agricultural products.

(c)   Comparative advantage is influenced by **research and development** both in technology and work organisation. The United States, because of the scale of its expenditure on research and development over many years, tends to have comparative advantage in goods embodying advanced technology.

(d)   A country should expect the **make-up of its exports and imports to change** over time. Thus the UK has moved from being the leading country for manufactured goods to a situation where service industries now play a leading role.

While the principles of free international trade are not universally accepted free trade brings greater competition and a threat to existing producers.

GMN viewpoint

While more and more companies are competing in the world market place, most of them tend to focus on the developed markets of North America, Europe and Japan. A vast majority (86%) of the world's population resides in countries where GDP is less than $10,000 per head. Such emerging markets such as the so-called BRIC countries (Brazil, Russia, India and China) offer tremendous marketing opportunities if the offering is presented correctly. BRIC countries are not the only emerging markets and there are other Big Emerging Markets (BEMs) such as Mexico, Poland, Turkey and parts of Africa.

Some **key factors** differentiating domestic and international marketing are outlined in the table below.

	Domestic	International
Social and cultural factors	Relatively homogeneous market 'Rules of the game' understood Similar purchasing habits	Fragmented, diverse markets Rules diverse, changeable and unclear Diverse purchasing habits
Economic factors	National price Uniform financial climate Stable business environment	Diverse national prices Variety of financial climates, ranging from very conservative to highly inflationary Multiple business environments, some unstable
Competitive factors	Competitors' products, prices, costs and plans usually known	Many more competitors, but little information about their strategies
Political legal factors	Relative freedom from government interference Political factors relatively unimportant	Involvement in national economic plans Political factors often significant
Technological factors	Use of standard production and measurement systems	Training of foreign personnel to operate and maintain equipment

Using the PESTLE framework above undertake an environmental audit for your own organisation or an organisation of your choice by comparing your own country with a country the organisation is considering entering. You should consider comparing and contrasting your analysis with your study buddy.

2 Motives and triggers of internationalisation

2.1 Internationalisation motives

Definition

The fundamental reasons – proactive and reactive – for **internationalisation**.

Reasons for internationalisation can be split into proactive and reactive motives. These include the following:

Proactive motives	Reactive motives
• Profit and growth objectives	• Competitive pressures
• Managerial desire, drive and enthusiasm	• Domestic market saturation or low growth
• Technological competence/unique product	• Intense competition in domestic markets
• Foreign market opportunities	• Unexpected foreign orders
• Economies of scale	• Extend sales of seasonal products
• Tax benefits	• Proximity to foreign markets/psychic distance
• The market is global	• Extend life cycle of maturing/declining products
• Identifying cheaper sources of materials and labour	• Competition have entered foreign markets
• Competition are not performing well in foreign markets	• A credible route to market has been identified through networking with others
• A global strategy will enhance reputation in the eyes of key stakeholders	

Adapted from Hollensen, S. (2007).

In terms of **competitive strategy**, global activities can be justified on the following grounds as outlined by Michael Porter in *Competitive Strategy* (1998).

(a) **Cost leadership**. The domestic market may be too small for the firm to reap the economies of scale which may be necessary for a cost leadership strategy.

(b) **Differentiation**. A differentiated product may appeal to a number of different overseas markets.

(c) **Focus**. Oddly enough, global marketing can be used to implement a focus strategy, which is a concentration on a core segment of consumers. The segment does not have to be defined in national or ethnic terms, but in terms of wealth, life-style and so forth. Haute couture fashion, perfumes and other luxury goods may appeal to people of a similar income bracket the world over.

(d) To **pre-empt competition** from overseas producers. World trade is becoming liberalised. Firms can expect incoming competition. They might have no alternative to competing abroad.

2.2 Internationalisation triggers

> **Definition**
>
> Internal or external events taking place to initiate **internationalisation.**

For internationalisation to take place someone or something inside or outside the organisation must initiate the process and carry it through to implementation. These triggers include the following:

Internal triggers	
Perceptive management	• Management make it their business to become knowledgeable about other markets
	• Open-minded to new opportunities
	• Foreign travel identifies new opportunities
Specific internal events	• New employee joins the organisation
	• Reduced sales forces organisation to look at other markets
	• Foreign enquiry

External triggers	
Market demand	• Growth in foreign markets
Competing organisations	• Knowledge and information about competitive activities
Trade and membership associations	• Attending a Global Marketing Network event about the growth prospects in India might trigger initiation
Outside experts	• Government bodies, chambers of commerce, consultancies and financial institutions might all be sources of information and funding that encourage internationalisation
Investor pressure	• The power of investors and their requirements might encourage internationalisation

> **Assessment advice**
>
> It will be useful to start considering the triggers and motives that are relevant to the organisation you are investigating for your assessment. Being aware of these factors will help you put your recommendations into context.

3 The development of the global marketing concept

3.1 Overview

Global marketing consists of finding and satisfying global customer needs better than the competition, and of coordinating marketing activities within the constraints of the global environment (Hollensen, 2007).

The response of the organisation to the global environment depends greatly on the management's assumptions and beliefs, both conscious and unconscious, about the nature of doing business around the world. This worldview can be defined as the EPRG Framework (Hollensen, 2007).

GMN Expert – Svend Hollensen

Global Marketing Network's preferred definition is from Svend Hollensen, a Global Marketing Network Global Advisory Council Member and Academic Advisor to this module.

Svend Hollensen has defined the EPRG Framework as an easy way of summarising the four orientations that an organisation can take in the face of globalisation. This is the preferred GMN definition:

1 **Ethnocentric**: home country of the organisation is superior and the needs of the home country are most important
2 **Polycentric**: each country is unique and therefore each country should be targeted in a different way
3 **Regiocentric**: the world consists of regions. The organisation tries to standardise its marketing programming within regions, but not across them
4 **Geocentric (global)**: the world is getting smaller. The organisation may offer global product concepts but with local adaptation (ie as IBM famously said "Think Global, Act Local")

Consider the orientation of your organisation and the strengths or limitations it provides in pursuing a successful global marketing strategy.

3.2 Defining global marketing

There are many definitions that refer to global marketing or international marketing or international business. All of the are equally valid and worth of investigation.

Hollensen defines global marketing as *"the organisation's commitment to coordinate its marketing activities across national boundaries in order to find and satisfy global customer needs better than the competition"* (Hollensen, 2007).

Let's break this definition down into its four constituent parts:

- **Coordinate its marketing activities** – integrating existing and new marketing strategies and implementing them across global markets, involving a degree of centralisation, delegation, standardisation and adaptation to the local market.

- **Find global customer needs** – carrying out international marketing analysis and analysing the macro and micro environment, identifying similarities and differences across countries and regions.

- **Satisfy global customers** – adapting distribution, product, pricing and communication strategies, adapting business models across different countries and regions.

- **Being better than the competition** – assessing, monitoring, evaluating and responding to global competition through delivering better value at lowest total cost, lower prices, securing superior distribution networks, creating communication strategies that resonate more closely with customers within different countries and regions.

3.3 Standardisation and adaptation

The issue that all organisations have to contend with in undertaking a global strategy is whether to globally standardise the marketing activity or adapt it to address local differences and expectations.

At one extreme this leads to the concept of a multi-domestic approach in which the organisation has a completely different strategy for every different market it enters. At the other extreme there is a global approach in which every aspect of the marketing activity is standardised in all countries.

In practice organisations will adopt a combination of standardisation and adaption approaches, globalising some elements of the programme and localising others.

The strategic approaches therefore broadly available to organisations are multi-domestic, global or regional, a third strategic approach where largely standardised marketing strategies are implemented across a specific region of the world.

The largest and most complex organisations in the world tend to use a combination of all of these strategies. In such cases the choice of strategy is influenced significantly by the supply chain decisions relating to the location of operations and outsourcing arrangements.

Zou and Cavusgil (2002) identified three perspectives on global marketing strategy: the global marketing perspective, the configuration perspective and the integration perspective. The global organisation locates its supply chain where it can be carried out efficiently.

Components are then transferred to another country for assembly, before the highly standardised finished goods are exported to countries where they are largely supported by standardised marketing.

Global organisations then integrate their activities and competitive moves to enable them to become major players in all major markets. Increased outsourcing makes all this more challenging as success ultimately is determined by customers' needs being met on time, profitably.

3.4 Globally standardised strategy

An organisation adopting such an approach makes no distinction between domestic and foreign market opportunities. Its perception is that markets are similar and that a single approach is required to satisfy each of the markets that it enters.

It enters markets that are consistent and supportive of its short term strategic objectives and goals. An organisation with such a focus formulates a long term strategy for the organisation as a whole and then coordinates the strategies of local subsidiaries to support this.

Meffet and Bolz (1993) describe the globalisation push and pull factors in terms of both the marketing programmes and the marketing processes.

Globalisation pull factors	Globalisation push factors
Globalisation of markets	**Globalisation of industries**
• Homogenisation of demand	• R&D expenses
• Global market segments	• Reduce payback cycles
• Globally active customers	• Experience curve effects
Marketing standardisation	**Globalisation of competitors**
• Programme standardisation	• Market interdependence
• Process standardisation	• Global competitors
	• Cross subsidisation

GMN viewpoint

Do bear in mind that the global business environment has changed considerably, and is continually changing since this model was created – many barriers to standardisation having been removed or reduced. Also globalisation effects such as economies of scale and high costs of innovation have become more significant drivers of standardisation. At the same time market fragmentation has increased.

In practice global firms have to strike a balance between the relative advantages of standardisation and adaptation to local tastes. So for example global food retailers such as McDonald's, KFC, Pret a Manger and Starbucks adapt their traditional global products to the tastes of different cultures and are also recognising the need for greater variety on their menus. They also recognise the need to adapt their pricing and promotional strategies to ensure they meet the requirements of the local market.

3.5 The concept of 'globalisation'

Some argue that global marketing organisations are rare. Industry structures change, foreign markets are culturally diverse, and the transformations brought about by developments in information technology mean that the world market is in a state of turbulence.

This leads on to the question of **market convergence** – how likely is it that consumers' tastes and preferences may converge? On the face of it, there is no reason why they should not, but in practice this is unlikely. Someone who lives in a very hot country is never going to want many sweaters, or be likely to go cycling in the midday sun just for the sake of getting some exercise.

Convergence theories do have strong anecdotal support. The average French high school student appears very similar to American students of the same age (clothing, eating and entertainment preferences). Take a student from Nigeria and compare him to one from Finland, however, and the story is likely to be different.

Going global can add to the value chain. Integrating the supply chain across several countries can lead to considerable cost savings. Increased levels of customer service allied with these cost savings can lead to a dominant market position. This needs an organisational culture that is aware of the supply chain and customer service whatever the country (or countries) of operation.

However it is widely recognised that a successful global marketing strategy will only succeed in today's competitive market by thinking globally and acting locally. This is known as the 'globalisation framework' (Hollensen, 2007). This is achieved through a dynamic and respectful interdependence between the corporate headquarters and country or regional subsidiaries and operations.

Organisations that follow such a strategy recognise the need to adapt their business model and to be flexible to the needs of the local market, while at the same time maintaining the efficiencies and maximising the value of a global process in order to ensure innovations created in one local market can be transferred and shared across the organisation as appropriate.

This, at times, causes conflicts and tensions between the local operation and the organisational HQ which must be managed effectively. The lack of trust caused through cultural distance can create resistance, friction and misunderstandings and management must be relentless in pursuing a knowledge of cross-cultural issues and implementing and transferring that knowledge across the organisation to maintain a sustainable competitive advantage.

Global case study

How Starbucks Colonised the World

With $7.8 billion (£4 billion) in annual revenues, 40m customers and 13,000 outlets, Starbucks now owns its market like few other companies in recent memory.

Given the chain's breakneck international expansion and ability to reshape coffee-drinking habits the world over, it is a global institution. It's no stretch to say it has changed the modern world.

In 1996, Starbucks unveiled its first Japanese outlet, in Tokyo's Ginza. At the opening ceremony, Shinto priests blessed the new store and prayed for harmony between Starbucks and the forces of nature, while Schultz and other executives sipped sake and washed their hands in spring water. When Schultz and Howard Behar, the Starbucks president, saw a line of 100 excited Japanese coffee fanatics waiting for their first taste of Starbucks, they cried.

Japanese consumers have long adored American brands, but the company's success there was far from preordained. In fact, when Starbucks was still mulling its Japanese debut, a consulting firm advised against it. Although the Japanese were already drinking impressive quantities of coffee by the mid 1990s – both from the cramped kis-saten shops that had long dotted the urban landscape and from the pervasive vending machines – the consultants believed Starbucks would clash with social norms.

Developing Global Marketing Strategy

The chain's no-smoking rule would alienate young people, they thought, and the etiquette-obsessed Japanese would never be seen drinking coffee in public.

Despite this, Starbucks made just a few, slight changes to its formula, such as concocting a green tea frappuccino, and offering smaller drinks and pastries to conform with local preferences. Then it began the deluge of outlets. And here's the crucial part: the Japanese bent over backwards to accommodate Starbucks, not the other way round.

Sometimes the company fundamentally alters its operations for a new market – as it did in Saudi Arabia, where all stores are segregated into "men-only" and "family" sections – in accordance with the country's rigid separation of the sexes in public.

After a warning from the Saudi religious police, the Muttawa, Starbucks also removed the scandalous, bare-breasted mermaid from its logo and replaced her with a simple crown. The Muttawa later had a change of heart, and the company reinstated the original logo.

Mostly, however, the changes are minor. The company can get away with this for a reason that might upset opponents of globalisation: international customers crave the full American Starbucks treatment, caramel syrup and all.

"The local consumer in all these countries wants the 'authentic' Starbucks experience," said Schultz. *"They don't want it watered down for Mexico or watered down for China. They want Starbucks."*

Plus, international customers often take a liking to Starbucks for the very same reasons Americans did. In Mexico City, Starbucks stores attract the young, the beautiful and the affluent. Consumers in Jakarta proclaim their pride in being able to drink the same cup of coffee as the country's wealthiest people, even though a frappuccino costs more there than the average Indonesian factory worker earns in a day. Kuwaitis like the dating scene, especially since the sexes almost never get to intermingle in public otherwise.

The Starbucks coffee concept has even taken off in war-torn Afghanistan, where a knock-off "Starbucks" in Kandahar is a community hub and a financial success, selling 500 cups of coffee a day.

The French appear to be embracing Starbucks even as they make a show of despising it. Starbucks made its French debut in January 2004, near the peak of American-French tension over the Iraq war. Schultz tried to reduce the friction by portraying his company as a bridge between the two nations, announcing that Starbucks had come *"to share our interpretation of coffee"*.

Already, you can find the chain in 37 countries, including unexpected places like Oman, Qatar, Chile and Cyprus. Seoul, South Korea, is home to a 200-seat café that spans five full storeys – the largest Starbucks in the world.

Just as McDonald's has transformed eating habits around the world by offering quick and cheap meals, Starbucks tends to make permanent cultural alterations as well.

Adapted from http://business.timesonline.co.uk/tol/business/industry_sectors/leisure/article3381092.ece

Accessed 2 April 2010.

3.6 The underlying forces for 'global coordination/integration' and market responsiveness

The terms global strategy and 'globalisation' have been introduced to reflect and combine two dimensions.

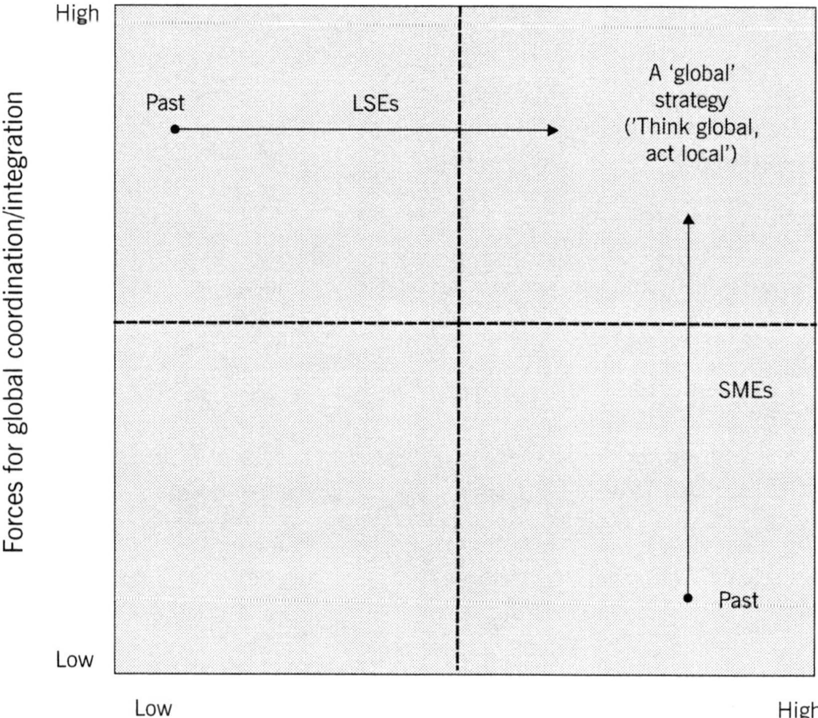

Forces for national market responsiveness

To achieve a 'global' strategy is to recognise the importance of creating a global integrated strategy that requires local adaptations/market responsiveness. In this way 'globalisation' tries to optimise the balance between standardisation and adaptation of the organisation's marketing activities (Svensson, 2001, 2002).

In the movement towards an integrated global marketing strategy greater importance will be attached to transnational similarities for target markets across national borders and less on cross-national differences.

Forces for global coordination/integration	Forces for market responsiveness
Removal of trade barriers (deregulation)	Cultural differences
Global accounts/customers	Regionalism/protectionism
Relationship management/network based organisation	Trend away from globalisation
Standardised worldwide technology	
Worldwide markets	
Global village	
Worldwide communication	
Global cost drivers	

Adapted from Sheth and Pravatiyar (2001) and Segal-Horn (2002).

3.7 Regional strategy

For many organisations regionalisation represents a more manageable compromise between the extremes of standardisation and globalisation. The key to developing effective regional strategies is in deciding what makes the region distinctive and in what ways the marketing strategy for one region should or needs to be

different from the others. Another critical success factor is in observing trends that pose threats or opportunities.

While regional strategies are built around the common elements within a region, in practice the differences within the region are still widespread.

3.8 The transnational strategy perspective

Many multinational organisations have a wide range of products and services, some of which might be suited to 'global' development. The successful exploitation of these opportunities requires a more flexible approach to strategic development and implementation which may involve a number of partners in licensing, franchising, joint ventures and strategic alliances as well as wholly owned companies.

Organisations with a transnational perspective aim to integrate diverse assets, resources and people into operating units around the world with the aim of building three main strategic capabilities to achieve global competitive advantage:

- Global scale efficiency and competitiveness
- National level responsiveness and flexibility
- Cross-market capacity to leverage learning on a worldwide basis

In such organisations the ownership of the operations becomes less clear in terms of the origin of where the product was made, what the domestic nationality is of the service or product provider, or which organisation produces the product or service. Such characteristics of these organisations include the following features:

- Some functions (such as marketing) being centralised, others having a substantial degree of independence

- A strong corporate identity and messages

- Strategic alliances with partners in order to carry out certain activities where partners add lasting benefit to its consumers

- Simple and complex product and market polices in equal measure, which may be independent or interdependent

- Customer segments that are specific and unique to a cross-functional niche market so that segments are valid across borders

- Maintaining and building meaningful and added value relationships in the supply chain

- Working closely with internal and external stakeholders in such a way that ensures the overall values and positioning of the organisation is maintained in all markets it operates within.

Activity 3

Consider the application of the above approaches and perspectives to internationlisation to the organisation you are investigating. What best describes the approach to internationalisation being undertaken? What apprach would be most suitable? Why?

4 The evolution of global marketing strategy

4.1 Overview

Here we present a number of different theoretical approaches to global marketing and the emergent issues gaining significance.

4.2 The Uppsala model

The Uppsala model (Johanson and Vahlne, 1977) identifies four different modes of entering a global market, where the successive stages represent higher degrees of involvement and commitment:

Stage 1:   No regular expert activities

Stage 2:   Export via independent representatives

Stage 3:   Establishment of a foreign sales subsidiary

Stage 4:   Foreign production/manufacturing units

Various criticisms have been put forward about this model:

- It is too deterministic

- Does not take into full account the interdependencies between different country markets

- The pace of expansion of new entrants has seen them leapfrog certain stages

- Organisations now have quicker and easier access to knowledge about doing business globally

- The world has become more homogenous and consequently the psychic distance between countries has decreased

- The number of people with experience of doing business globally has increased

- Information technologies and the internet has made it easier for an organisation to become familiar with foreign markets; thus making a leapfrog strategy more realistic

4.3 The network model

> **Definition**
>
> In a business network 'actors' are autonomous and linked to each other through relationships which are flexible and may alter accordingly to rapid changes in the environment. The 'glue' that keeps the relationships together are based on technical, economic, legal and personal ties that lead to mutual commercial benefit.
>
> The **network model** enables the relationships in one organisation in a domestic network to be used as bridges to other networks in other countries.

The network model is adopted by organisations where coordination can give strong gains for all parties and where conditions are changing rapidly. It relies more on a relationship-based approach than a transactional based approach where there is mutual respect and recognition between the parties for the value they bring.

The structure and management of such an organisation cannot be easily observed by a competitor and creates a barrier to entry to new entrants. The development of the organisation is based on the integration

of the resources in the network; the stronger the partner, the stronger the sense of a shared vision, the greater the opportunity for sustainable competitive advantage.

Entering a network requires a commitment to partnership. It may require the organisation to change or adapt its existing business practices or business model.

Networks in one country may well extend beyond country borders to create global networks. This may involve certain aspects of the organisation's operation being headquartered in one country, products being developed for one domestic country being adapted for other countries and so on. The role of the organisation is to maintain, build and co-ordinate the activities of the network to ensure maximum value is leveraged at lowest total cost.

Activity 4

Consider the advantages and limitations of each theory of internationalisation and which best support your organisation's global marketing aspirations?

4.4 The transaction cost analysis (TCA) model

Definition

Transaction Cost Analysis (TCA). The foundation for this model was made by Coase (1937). His view was that an organisation will need to expand until the cost of organising an extra transaction within the organisation will become equal to the cost of carrying out the same transaction by means of an exchange on the open market.

TCA predicts that an organisation will perform those activities it can undertake at lower cost through establishing an internal management control and implementation system while relying on the market for activities in which outside agencies have a cost advantage.

Transaction costs can be divided into ex ante costs and ex post costs:

Ex ante costs	Ex post costs
Search costs: costs of gathering information to identify and evaluate potential export intermediaries	Monitoring costs: costs of monitoring the agreement
Contracting costs: costs of negotiation and gaining agreement	Enforcement costs: costs associated with sanctioning a trading partner who does not perform in accordance with agreement

A fundamental assumption of TCA is that organisations will attempt to minimise the combination of these combined costs.

Therefore, for example, if the transaction costs through external agencies are higher than control through an internal system the organisation should seek to internalise as much as possible in the form of its own subsidiaries.

4.5 Global market selection and expansion

In selecting a market the organisation will need to consider the following issues:

The organisation

- The degree of internationalisation and overseas experience
- Size and amount of experienced resources
- Type of industry and nature of the organisation
- Global goals and aspirations of the organisation
- Existing network and relationships

The environment

- Global industry structure
- Degree of internationalisation of the markets
- Host country – market potential, competition, psychic/geographic distance, market similarity

Once these issues have been considered it will lead to the development of global market segmentation. On the basis of this it will be possible to select the global market or markets that are most attractive to enter.

Step 1   Define global market segmentation criteria

Step 2   Screen potential markets and countries against specific criteria to indentify qualifying countries

Step 3   Develop sub-segments in each qualified country and across countries

An example of this systematic approach is provided in the following diagram.

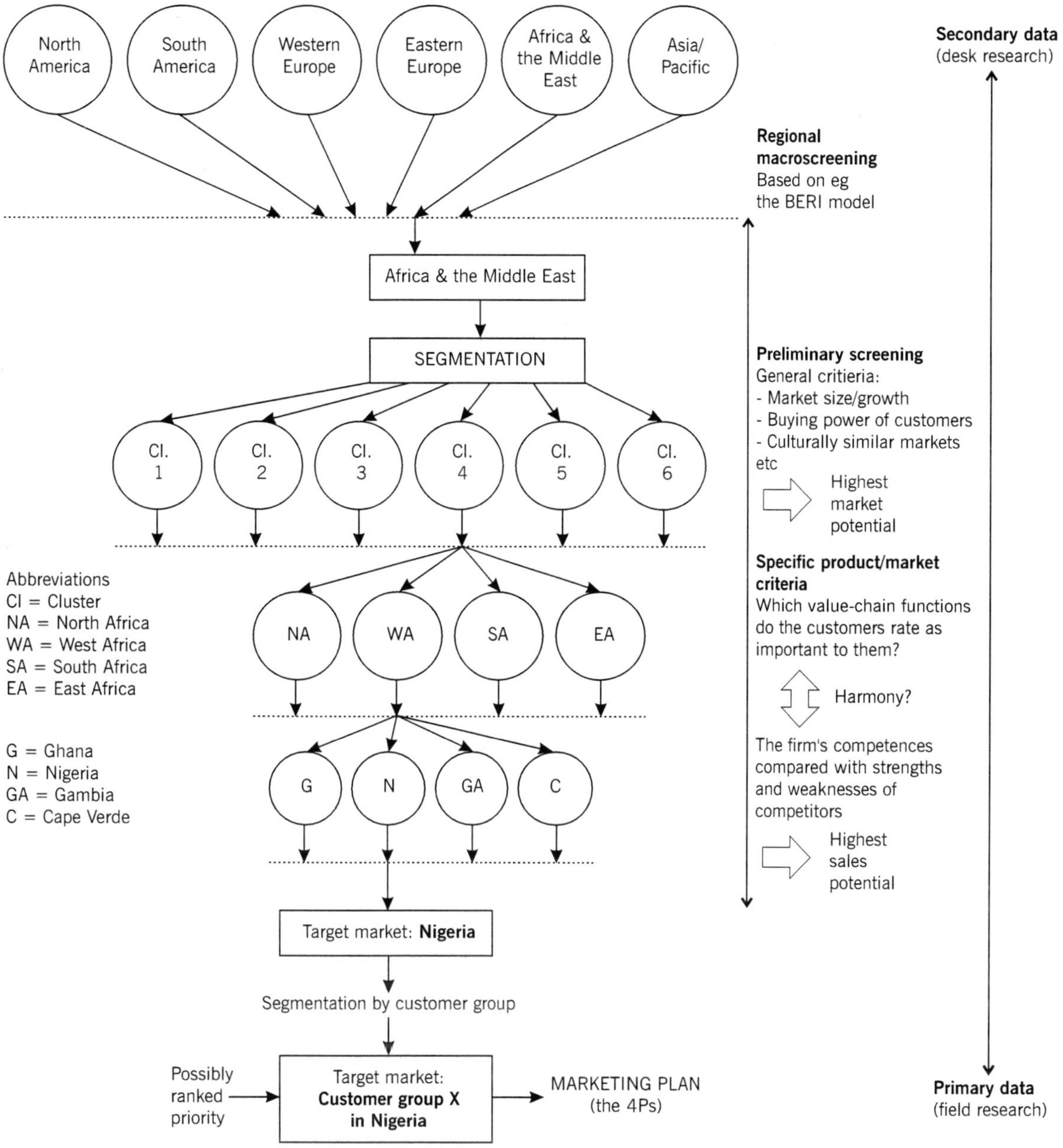

North America · South America · Western Europe · Eastern Europe · Africa & the Middle East · Asia/ Pacific

Regional macroscreening
Based on eg the BERI model

Africa & the Middle East

SEGMENTATION

Cl. 1 · Cl. 2 · Cl. 3 · Cl. 4 · Cl. 5 · Cl. 6

Preliminary screening
General critieria:
- Market size/growth
- Buying power of customers
- Culturally similar markets etc

Highest market potential

Abbreviations
CI = Cluster
NA = North Africa
WA = West Africa
SA = South Africa
EA = East Africa

G = Ghana
N = Nigeria
GA = Gambia
C = Cape Verde

NA · WA · SA · EA

Specific product/market criteria
Which value-chain functions do the customers rate as important to them?

Harmony?

The firm's competences compared with strengths and weaknesses of competitors

Highest sales potential

G · N · GA · C

Target market: **Nigeria**

Segmentation by customer group

Possibly ranked priority →

Target market:
Customer group X in Nigeria

→ MARKETING PLAN (the 4Ps)

Secondary data (desk research)

Primary data (field research)

Adapted from Hollensen (2007)

An organisation will need to decide whether:

1. to enter global markets on an incremental or experiential basis, entering first a single key market to build up experience in global marketing and then entering other markets one after the other

    or

2. enter a number of markets simultaneously in order to leverage its core competence and resources rapidly across a broader market base

Incremental benefits	Simultaneous benefits
Organisation gradually builds confidence	Seize emerging global opportunity quickly
Familiarity and information gained step-by-step	Early market penetration
Ideal where organisation is small and lacks resources	Build rapid experience
	Achieve economies of scale in production and marketing
	But requires substantial financial resources and is higher risk

4.6 Developing countries and emerging markets

It is widely predicted that the majority of economic growth over the next two decades will come from developing and newly industrialised countries. These are sometimes referred to as Big Emerging Markets (BEMs).

BEMs share a number of important traits:

- Geographically large
- Significant populations
- Represent sizable markets for a wide range of products
- Strong rates of growth / potential for significant growth
- Undertaken significant programmes of economic reform
- Are of major political importance in their region
- Are "regional economic drivers"
- Will engender further expansion in neighbouring markets as they grow
- Have a young demographic with well educated, aspirational young people who are connected to the global village

Brazil, Russia, India and China (BRIC nations) as well as Mexico, Poland, Turkey and South Africa are among those regarded as BEMs. Much of their expected growth is expected in industrial sectors such as IRT, environmental technology, energy technology, healthcare technology and financial services.

Clearly marketing activity will need to be tailored for BEMs to ensure it meets the needs of the target market and is conducive to the local environment. As the country develops, market behaviour changes and eventually groups of consumers with common tastes and needs arise.

Emerging markets will be the growth areas of the 21st century and it is vital that marketers identify, analyse and evaluate their importance to their organisation and industry sector.

GMN viewpoint

Martyn Mensah, GMN Ghana Honorary President

Speaking at World Marketing Forum, Accra, Ghana; 17 February 2010

It is entirely appropriate that GMN is setting out its stall on the continent of Africa, sometimes described as the last frontier. Ours is indeed a continent pregnant with opportunity and therefore, now more than ever is the era of the African marketing professional; an individual with a solid understanding of what marketing is and how it contributes value to the entire business of business; an ambitious professional who sets very high standards and who exercises a relentless focus on the changing consumer and relates it to the evolution of consumers in other countries and in other times. The means to seizing the moment and making the most of the African opportunity lies in adopting a global and professional posture, in seeking to be leaders in all spheres of endeavour.

As a continent, we have tried everything to get ourselves unto a growth trajectory – aid, debt, foreign direct investment, debt forgiveness, structural adjustment … and yet we are where we are. May I humbly propose that now is the time for a marketing led renaissance of the Continent. Such an undertaking can only succeed with the growth in skill, knowledge and numbers of a cohort of professional African marketers.

4.7 The impact of technology on global marketing strategy development

For some organisations, such as Dell, Amazon and Google for example, their global marketing strategy is inextricably linked to technology.

Increasingly technological competence and capability as well as understanding the competitive market position and gaining in-depth customer insights is becoming a critical success factor.

The Internet and pressure groups mean that there is now a greater transparency about companies and their activities. Reputation management through having a sound approach to supplier selection, the environment and ethical issues will have a big impact on how the brand is perceived.

4.7.1 Product and service management

- Technology integrates global members of the supply chain
- Technology also speeds up innovation
- Creating the ability to offer global product platforms as well as increased customisation

4.7.2 Pricing

- Increasing price transparency
- Costs being driven down
- More difficult to operate price differentials across borders
- Technology can enable price customisation

4.7.3 Channel management

- Encourages removal of intermediaries in the supply chain
- Lower transactional costs
- Manufacturer or producer has stronger control over the market and closer relationship with the customer
- Manufacturer or producer can share more market knowledge and achieve closer cooperation with its intermediaries

4.7.4 Communications

- More targeted and interactive communication activity

4.7.5 Control, evaluation and learning

- Can collect, transfer and analyse vast amounts of data from anywhere
- Better financial management and control
- Most effective when supported by an organisational culture and structure that encourages the sharing of knowledge across its worldwide operation

4.8 The rise of the 'born globals'

In recent years research has identified an increasing number of organisations that do not follow traditional patterns in the development process of globalisation. These are known as 'born globals'.

Characteristics of born globals
• Many activities coordinated across many countries
• SMEs with less than 500 employees and annual sales under £100m
• Reliance on cutting-edge technology in the development of relatively unique product or process innovations
• Managed by entrepreneurial visionaries who view the word as a single, borderless marketplace from the time of the organisation's founding
• Small, technology-oriented organisations that operate in global markets from the earliest days of their establishment

- Background of the decision maker(s) largely influences the strategy

- Network based approach to sales and marketing through a partnership-based approach that complements their own competences

- Collaboration and partnerships are established to facilitate rapid growth and global development

Adapted from Hollensen, 2007

A number of factors have given rise to the emergence of the 'born global' and several trends explain their increasing importance and why such organisations can effectively implement global strategies:

Trends	
Increasing role of niche markets	• Growing demand for specialised or customised products. Smaller organisations are supplying products that occupy a relatively narrow niche and become experts in that niche over their larger competitors
Advances in process/technology production	• New technology enables smaller organisations to compete on a comparable footing with large multinationals in the production and delivery of sophisticated products and services
Flexibility of SMEs/born globals'	• Quicker, more flexible, more adaptable, ensures they can adapt to consumer tastes and requirements more efficiently
Global networks	• Success in global trade is increasingly being achieved through establishing partnerships and long term alliances with organisations to increase value to the market at lower total cost
Advances and speed in information technology	• The expansion of the internet and digital technologies worldwide has ensured that rapid decision making can now be taken around the world. Small organisations that can harness this technology with rapid decision making are able to obtain an advantage over their larger, slower, more bureaucratic competitors.

4.9 The anti-globalisation challenge

Recent years have seen a backlash against the notion of globalisation. Multinational companies such as Nike and Shell have been the target of attacks on their policies, including but not limited to workers' pay, in developing countries. Ethical issues have become globalised, and companies are having to operate with higher standards.

Many countries benefit from globalisation, but it is also true that many do not. A few developing countries attracting the greatest share of investment (such as China, Brazil, India and the Philippines) have increased their trade, with GDP per head increasing often at a higher rate than in richer countries. However, billions of people live in countries that have become less rather than more globalised (Pakistan and much of Africa, for example). Economic growth in such areas has been stagnant, and poverty has risen. World institutions such as the IMF and World Bank have significant influence on development of world trade and the restructuring of struggling economies. However these organisations are not universally admired, with many blaming them for the financial difficulties faced by developing countries struggling with huge debts.

Following several years of research, Naomi Klein completed a book *No Logo* in 2000 that criticises the business practices of large multinational corporations as well as the policies of international organisations like the World Trade Organization (WTO). The timing of this book was perfect as it was published shortly after the 1999 WTO summit in Seattle, where a mass protest rally by anti-globalisation activists turned into a riot. *No Logo* expresses powerfully the anger that anti-globalisation protesters feel about what is going on in the world. This book immediately became a bestseller and it has been very influential. As a consequence, Naomi Klein has emerged as an intellectual leader in the anti-globalisation movement.

Visit her website at: http://www.naomiklein.org/main

1 Consider the impact of international trade

- There is no global marketing without international trade.

- Larger profits are achievable as a result of volume sales, specialisation, segmentation of wider markets and wider resources sourced.

2 Understand the motives and triggers of internationalisation

- Proactive and reactive motives exist.

- Triggers are required for internationalisation to be possible.

3 Evaluate the development of the global marketing concept

- The ERPG Framework summarises the four orientations: ethnocentric, polycentric, regiocentric or geocentric.

- Standardisation or adaption is a key consideration.

- Market convergence refers to consumers tastes and preferences becoming increasingly similar.

- Transnational and regional strategies are options for global marketing.

4 Understand the evolution of global marketing strategy

- Various approaches and models of global marketing exist including the Uppsala model, the network model and transaction cost analysis.

- Born global organisations do not follow traditional processes towards globalisation but launch in many countries.

1   This answer will largely depend upon your organisation. The PESTLE framework will form a reference to enable you to compare and contrast your country with another.

2   This will depend on your own research.

3   This answer will largely depend upon your organisation and its preferred orientation. Consider Hollensen's EPRG Framework. Which is your organisation's preferred orientation? Consider whether it needs to adapt its orientation in light of changing circumstances in the PESTLE and competitive environment.

4   This answer will largely depend upon your organisation. There are advantages and limitations associated with each theory of internationalisation.

Cateora, P., Graham, J., (2009). *International Marketing*. 14th edition, London: McGraw Hill.

Doole, I. & Lowe, R., (2008). *International Marketing Strategy Analysis, Development and Implementation*. 5th edition, London: Thomson Learning.

Hollensen, S., (2007). *Global Marketing*. 4th edition. London: Prentice Hall.

Johanson, K. and Vahlne, J. E., (1990). *The Mechanism of Internationalisation*: Almquist & Wiksell International.

Meffet and Bolz, (1993). In Halliburton and Hunerberg (EDS), *European Marketing Reading and Cases*, London: Addison-Welsey.

Segal-Horn, S., (2002). 'Global Firms: heroes or villains? How and why companies globalise'. *European Business Journal*.

Sheth, J. N. and Pravatiyar, A., (2001). 'The antecedents and consequences of integrated global marketing' *International Marketing Review*.

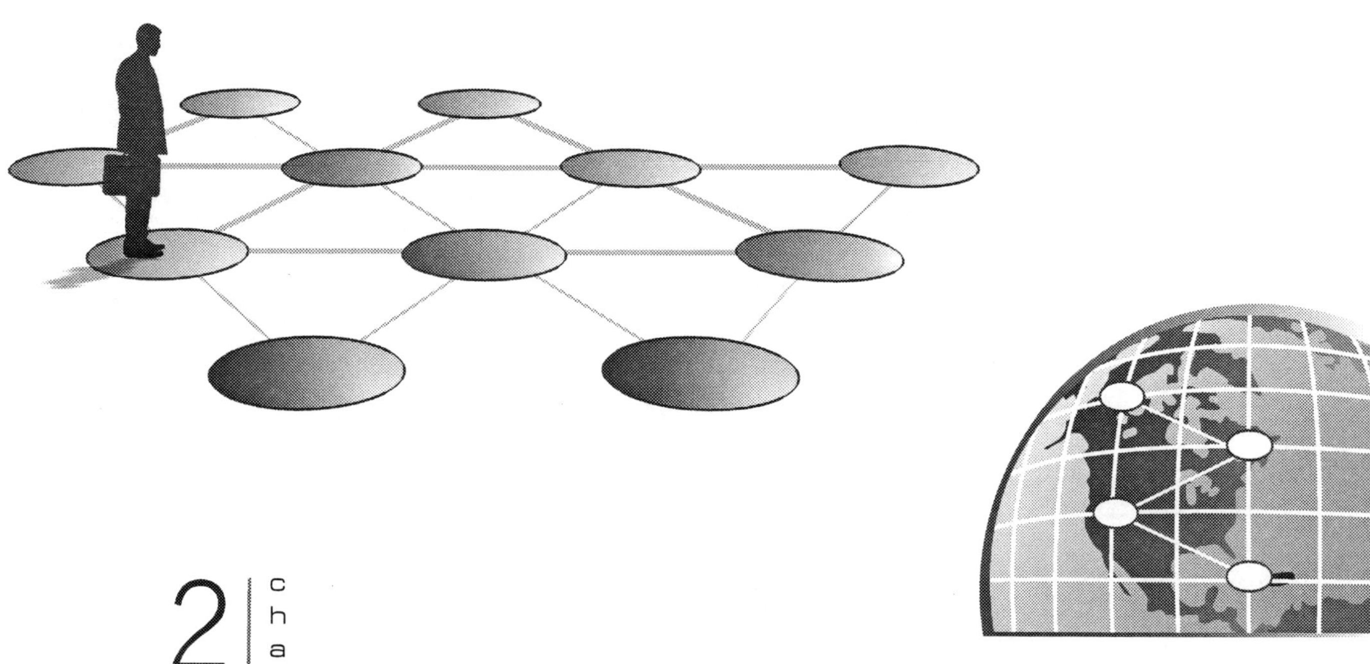

2 chapter

Understanding the political, legal and economic environment

Global marketing takes place within the framework of an international trading environment. If a marketer is to have the necessary capabilities to develop effective global marketing strategies some understanding of the parameters of the global trading environment within which they operate is required. In this chapter we examine the development of international trade in recent years. We explore the political, legal and economic factors and issues that the marketer faces in developing and implementing a successful global marketing strategy and the institutions that aim to influence international trade. We also explore the changing regional trade blocs and the implications that these have on trading around the world.

Contents

Chapter learning outcomes

By the end of this chapter you will be able to:

- Appreciate the importance of the political and legal environment
- Explore the impact of protectionism in global trade
- Understand the importance of economic structure and development
- Explore the concept of the balance of payments
- Understand the role of regional trading groups
- Understand the nature of international institutions and agreements
- Appreciate the impact of currency fluctuations and exchange rates

1 Setting the scene

This chapter focuses on the political, legal and economic complexities faced by the organisations competing globally which do not exist in the domestic market. Organisations need to understand the global marketing environment clearly in order to succeed.

Definition

SLEPT (or **PESTLE**) (social, legal, economic, political and technological) factors are all key issues in an organisation's domestic environment. The same factors apply when we are discussing an organisation's global exposure - they just become more complex.

Assessment advice

Undertaking a SLEPT (or PESTLE) analysis for the domestic environment and the environment of the country you are seeking to enter is a critical step in the development of and rationale for your global marketing strategy. As you work through this session you should consider the issues that your organisation faces, the opportunities these offer, the challenges they offer and the potential solutions.

2 The political and legal framework

At some time, most organisations engaged in global marketing suffer because of the political or legal structure of a country. There is political risk in engaging with every county but in general risk is lowest in countries which have a history of stability and consistency.

2.1 Political risk

Definition

Political risk is the possibility of turbulence (eg civil war, revolution, changes in government policy) in the political environment. It also encompasses risk from protectionism and government regulation.

The level of risk involved will depend on several factors.

- The attitudes of the country's government
- The product being traded
- The company wishing to trade

2.1.1 Political government of the country

The development of plans for global marketing will depend on the following factors.

(a)   The stability of the government. Rapid changes or political unrest make it difficult to estimate reactions to an importer or a foreign business.

(b)   Global relations. The government's attitude to the organisation's home government or country may affect trading relations.

(c)   The ideology of the government and its role in the economy will affect the way in which the company may be allowed to trade, and this might be embodied in legislation.

(d)   Informal relations between government officials and businesses are important in some countries. Cultivation of the right political contacts may be essential for decisions to be made in your favour.

2.1.2 The product

The nature of the goods or services being offered may affect the degree of interest which a government takes in a particular trading deal. Generally the more important the goods to the economy or the government, the more interest will be taken.

2.1.3 The organisation wishing to trade

The previous relations of the organisation (and its home country) with the host country can affect the risk to the company. Factors influencing the company's acceptability include the following.

- Relations between the company's home government and the overseas government
- Size of the company
- The past relations and reputation of the company in dealing with foreign governments
- The degree of local employment and autonomy of operations generated by the activity

2.1.4 Expropriation and other dangers

Political risk is still relevant with regard to overseas investment, especially in large infrastructure projects overseas. History contains dismal tales of investment projects that went wrong, and were expropriated (nationalised) by the local government.

(a)   Suspicion of foreign ownership is still rife, especially when prices are raised.

(b)   Opposition politicians can appeal to nationalism by claiming the government sold out to foreigners.

(c)   Governments might want to renegotiate a deal to get a better bargain, at a later date, thereby affecting return on investment.

In addition to expropriation, there are other dangers.

- Restrictions on profit repatriation (eg for currency reasons)
- Cronyism and corruption leading to unfair favouring of some companies over others
- Arbitrary changes in taxation
- Pressure group activity

There are many sources of data. The *Economist Intelligence Unit* offers assessment of risk. Management consultants can also be contacted. According to Jeannette Hennessey, (2002) companies should ask the following six questions.

1	How stable is the host country's political system?
2	How strong is the host government's commitment to specific rules of the game, such as ownership or contractual rights, given its ideology and power position?
3	How long is the government likely to remain in power?
4	If the present government is succeeded, how would the specific rules of the game change?
5	What would be the effects of any expected changes in the specific rules of the game?
6	In light of those effects, what decisions and actions should be taken now?

Coping with political risk

The approach taken depends on the degree of risk and the level of involvement.

Level of risk

High	• Keep low profile	• Contingency plans
	• Communicate via third parties	• Disinvest?
	• High level contacts	• Act for stability?
	• Short term deals	
	• Export credit insurance	
Low	*No need to worry*	*No need to worry - but monitor for increasing riskiness*
	Low	*High*

Level of involvement

Political risk analysis procedure

Hollensen (2007) outlines a procedure for analysing political risk and avoiding the common error of underestimating such risk at the level of the organisation. This involves three major steps:

Step 1   Assessing the issues of relevance to the organisation

Step 2   Assessing potential political events

Step 3   Addressing political risks through relationship building with:

    (a)   Government
    (b)   Customers
    (c)   Employees
    (d)   Local community

Further measures to reduce political risk are as follows.

- Use local partners with good contacts
- Vertical integration of activities over a number of different countries
- Local borrowing (although not a good idea in high inflation countries)
- Leasing rather than outright purchase of facilities in overseas markets
- Take out insurance

Activity 1

Undertake a Political Risk Analysis Procedure for your organisation for the countries you are considering entering. Discuss your findings with your colleagues or your Study Buddy and determine the extent to which these can be overcome.

2.2 Legal factors

In global markets we are interested in legislation which may affect an organisation's trade with a particular country.

- Domestic legal system
- Structure of company law
- Local laws

Legal implications extend far beyond the marketing mix. Each country may legislate on the following issues, and these may affect the marketer to a greater or lesser degree.

(a)   **Export and import controls** for political, environmental, or health and safety reasons. Such controls may not be overt but instead take the form of bureaucratic procedures designed to discourage global trade or protect home producers.

(b)   **Favourable trade status** for particular countries, eg EU membership, former Commonwealth countries.

(c)   **Monopolies and merger legislation**, which may be interpreted not only within a country but also across nations. Thus the acquisition of a company in country A, by company B, which both sell in country C may be seen as a monopolistic restraint of trade.

(d)   **Law of ownership**. Especially in developing countries, there may be legislation requiring local majority ownership of a organisation or its subsidiary in that country, for example.

(e)   **Taxation law** may be used to encourage or discourage particular import/export activities. For example generous tax incentives for inward investment may be offered.

(f)   **Acceptance of global trademark, copyright and patent conventions**. Not all countries recognise such global conventions.

(g)   Determination of minimum **technical standards** which the goods must meet, eg noise levels, contents and so on.

(h)   **Standardisation measures** such as packaging sizes.

(i)   **Pricing regulations**, including credit (eg, some countries require importers to deposit payment in advance and may require the price to be no lower than those of domestic competitors).

(j)   **Restrictions on promotional messages**, methods and media.

(k)   **Product liability**. Different countries have different rules regarding product liability (ie the manufacturer's/retailer's responsibility for defects in the product sold and/or injury caused). US juries are notoriously generous in this respect.

---

3 Protectionism in global trade

> **Definition**
>
> **Protectionism** is the discouraging of imports by raising tariff barriers, imposing quotas etc in order to favour local producers.

We will now identify the different forms of protectionist measures available to governments. Some governments seek to prevent the influence of global trade by making it harder to import from overseas.

- Tariffs or customs duties
- Non tariff barriers
- Import quotas and embargoes
- Subsidies for domestic producers
- Exchange controls
- Exchange rate policy

3.1 Tariffs or customs duties

A tariff is a tax on imports.

(a)   The importer is required to pay either a percentage of the value of the imported good (an *ad valorem* duty), or per unit of the good imported (a specific duty).

(b)   The government raises revenue and domestic producers may expand sales, but consumers pay higher prices if they buy imported goods. They may have to buy domestic goods of a lesser quality.

3.2 Non tariff barriers

3.2.1 Import quotas

Import quotas are restrictions on the quantity of a product allowed to be imported into a country.

(a)   The restrictions can be imposed by import licences (in which case the government gets additional revenue) or simply by granting the right to import only to certain producers.

(b)   Prices will rise because the supply of goods is artificially restricted. The consumer pays more while foreign producers benefit. Japanese firms may have benefited financially from quotas on their goods (eg cars) in the past, imposed by some European countries and the US. Quotas, like tariffs, are also likely to provoke retaliation.

3.2.2 Minimum local content rules

Related to quotas is a requirement that, to avoid tariffs or other restrictions, products should be made in the country or region in which they are sold. In the EU the product must be of a specified minimum local content (80% in the EU) to qualify as being 'home' or 'EU-made'. This is one of the reasons Japanese and Korean manufacturers have set up factories in Europe.

3.2.3 Minimum prices and anti-dumping action

Dumping is the sale of a product in an overseas market at a price lower than charged in the domestic market. Anti-dumping measures include establishing quotas, minimum prices or extra excise duties.

3.2.4 Embargoes

An embargo on imports from one particular country is a total ban, a zero quota. An embargo may have a political motive, and may deprive consumers at home of the supply of an important product.

3.2.5 Subsidies for domestic producers

An enormous range of government subsidies and assistance for exporters is offered, such as export credit guarantees (insurance against bad debts for overseas sales), financial help and assistance from government departments in promoting and selling products. The effect of these grants is to make unit production costs lower. These may give the domestic producer a cost advantage over foreign producers in export as well as domestic markets.

3.2.6 Exchange controls and exchange rate policy

Many countries have exchange control regulations designed to make it difficult for importers to obtain the currency they need to buy foreign goods. If a government allows its currency to depreciate, imports will become more expensive. Importers may cut their profit margins and keep prices at their original levels for a while, but sooner or later prices of imports will rise. A policy of exchange rate depreciation in this context is referred to as a competitive devaluation.

3.3 Unofficial non-tariff barriers

Some countries are accused of having unofficial barriers to trade, perpetrated by government. Here are some examples.

(a)   Quality and inspection procedures for imported products, adding to time and cost for the companies selling them

(b)   Packaging and labelling requirements may be rigorous, safety and performance standards difficult to satisfy and documentation procedures very laborious

(c)   Standards which are much easier for domestic manufacturers to adhere to

(d)   Restrictions over physical distribution

(e)   Toleration of anti-competitive practices at home

Non tariff trade barriers in detail

Formal trade restrictions	Administrative trade restrictions

A Non tariff import restrictions (Price related measures)

Surcharges at border

Port and statistical taxes

Non discriminatory excise

Taxes and registration charges

Discriminatory excise taxes

Government insurance requirements

Non discriminatory turnover taxes

Discriminatory turnover taxes

Import deposit

Variable levies

Consular fees

Stamp taxes

Various special taxes and surcharges

B Quantitative restrictions and similar specific trade limitations (Quantity-related measures)

Licensing regulations

Ceilings and quotas

Embargoes

Export restrictions and prohibitions

Foreign exchange and other monetary or financial controls

Government price setting and surveillance

Purchase and performance

Requirements

Restrictive business conditions

Discriminatory bilateral arrangements

Discriminatory regulations regarding countries of origin

Global cartels

Orderly marketing agreements

Various related regulations

C Discriminatory freight rates

D State participation in trade

Subsidies and other government support

Government trade, government monopolies and granting of concessions or licences

Laws and ordinances discouraging imports

Problems relating to general government policy

Government procurement

Tax relief, granting of credit and guarantees

Boycott

E Technical norms, standards and consumer protection regulations

Environmental emission standards

Health and safety regulations

Pharmaceutical control regulations

Product design regulations

Industrial standards

Size and weight regulations

Packing and labelling regulations

Package marking regulations

Regulations pertaining to use

Regulations for the protection of intellectual property

Trademark regulations

F Customs processing and other administrative regulations

Antidumping policy

Customs calculations bases

Formalities required by consular officials

Certification regulations

Administrative obstacles

Merchandise classification

Regulations regarding sample shipments return shipments, and re-exports

Countervailing duties and taxes

Appeal law

Emergency law

Consider the impact that protectionism can have on your organisation.

China leading global economy in recovery

China's industrial production and trade surplus posted robust double-digit gains in October, indicating a strengthening recovery in the world's third-largest economy. China, unlike the U.S. and other western industrial nations, has managed well in the advance of the global economic crisis. Such economic strength is also likely to increase pressure on policy makers to let the yuan appreciate as a hedge against inflation risks.

"For China, it is necessary and appropriate to allow the currency to be more flexible," Asian Development Bank President Haruhiko Kuroda said in an interview with Bloomberg. *"Crisis response by the Chinese authorities has been excellent,"* and *"they've brought about a very strong economic recovery,"* he added. As a result, production rose 16.1% year-over-year, the most since March 2008, according to China's state statistics bureau in Beijing. Meanwhile, retail sales gained an annual 16.2% in October. In addition, the trade surplus almost doubled from September, to $24 Bn, as the slide in exports eased to the slowest pace this year.

Chinese central bank policymakers have indicated they will improve the setting of the yuan's foreign exchange rate in a proactive, controlled and gradual manner and based on international capital flows and movements in major currencies.

Adapted from: http://globaleconomy.foreignpolicyblogs.com/

4 Economic structure and development

Economic factors affect the demand for, and the ability to acquire, goods and services. Even in lesser developed countries (see below) there often exists a wealthy elite who provide a significant demand for sophisticated consumer goods.

Countries generally have larger agricultural sectors in the early stages of economic development (for example India and Africa). As the economy develops, the manufacturing sector increases. The industrialised countries publish detailed statistics - the production of hundreds of industries may be recorded.

The correct classification of economic information is the subject of global standards. Industrial production, imports and exports are classified under standard headings. Many countries now follow the United Nations' Global Standard Industrial Classification (ISIC).

4.1 Level of economic development

Commonly, economists and marketers categorise countries into five broad types.

4.1.1 Classification of economic development

Generally each country can be classified under one of five headings.

- **Lesser developed country (LDC).** Relies heavily on primary industries (mining, agriculture, forestry, fishing) with low GDP per capita, and poorly developed infrastructure.

- **Early developed country (EDC).** Largely primary industry based, but with developing secondary (manufacturing) industrial sector. Low but growing GDP, developing infrastructure.

- **Semi developed country (SDC).** Significant secondary sector still growing. Rising affluence and education with the emergence of a 'middle class'. Developed infrastructure.

- **Fully developed country (FDC).** Primary sector accounts for little of the economy. Secondary sector still dominates, but major growth in tertiary (service) sector. Sophisticated infrastructure.

- **Former Eastern Bloc country (EBC).** May be any of the above, but the 'command economy' under communism has left a legacy that defies straightforward classification. For example, Russia, has most of the features of an SDC but lacks a developed infrastructure though it has a well educated middle class.

Each type then exhibits a fairly consistent pattern of demand for goods and services. Commonly used factors in classifying countries include the following.

(a) **Infrastructure** (eg the development of roads, transport, communications and energy distribution)

(b) **Education** and literacy levels

(c) **Ownership of durables** (for example, telephone, TV, fridge, etc as appropriate)

(d) **GDP** (Gross Domestic Product) per head. This, effectively, is the value of goods and services produced within the economy

All of the first three may be claimed to be dependent on GDP. GDP on a per capita basis, suitably adjusted for purchasing power, is probably the best single indicator of economic development. A danger in using GDP is that it considers only the average.

The distribution of wealth is critical in poor countries, where a market may exist among above average sections of the population.

4.2 Measuring levels of economic development

- Measures that may be used by the global marketer include the following.

    - GDP per head
    - Source of GDP (primary, secondary or tertiary sector based economy)
    - Living standards (ownership of key durables may be used as a surrogate measure)
    - Energy availability and usage
    - Education levels
    - Environmental issues are becoming more important since the Rio Summit

4.3 Identifying market size

The economic worth of a consumer market is based on some general factors.

- The **number of people** in the market
- Their **desire to own** the goods
- Their **need to own** goods
- Their **ability to purchase** the goods

Thus in measuring a market, the marketer will obtain information on the following, although they are often crude measures.

(a) **Population.** Its size, growth and age structure, household composition, urban v rural distribution. Household size and spatial distribution affect demand for many consumer goods.

(b) **Income.** GDP per head is a crude measure of wealth and account should also be taken of distribution of GDP among various social groups, and their purchasing power.

(c) **Consumption patterns.** The ownership of various goods and the consumption of consumables are indicators of potential demand.

(d)   **Debt and inflation**. A high level of debt in a country may indicate import controls (or their possible introduction) or weak currency and currency controls. Inflation may affect purchasing power. In either case ability to pay will be reduced.

(e)   **Physical environment**. Physical distance, climate and topography will affect demand in various ways. The availability of natural resources can directly affect demand for equipment and so on to exploit these resources.

(f)   **Foreign trade**. The trade relations of a country will affect the attitude towards foreign goods. Factors include economic relations (for example, a member of same economic group such as the EU) and balance of trade.

5 The balance of payments

> **Definition**
>
> The **balance of payments** is the statistical accounting record of all of a country's external transactions in a given period. The balance of payments account of any country records the revenues and payments during the course of a year from all economic transactions between its government, firms and residents, and the corresponding counterparts in those countries with which it trades.

Global trade can present countries with balance of payments problems arising from a persistent mismatch in the flows of exports and imports, resulting in **surpluses** or **deficits** in the value of goods and services exchanged.

The potential problems arising from persistent deficits are readily identified.

(a)   Since a balance of payments deficit represents a leakage of income from the national economy, there is a danger that economic growth will be retarded, unless internally generated growth can compensate for this.

(b)   A persistent deficit is likely to put downward pressure on the exchange rate as confidence in the currency is weakened and a demand for it falls.

(c)   A depreciating exchange rate will mean that the price of imports in domestic terms will be rising, putting pressure on domestic inflation with consequent knock-on effects for wage demands and unemployment.

The potential problems arising from persistent surpluses are not so readily identified. Nevertheless, problems can emerge.

(a)   A balance of payments surplus represents an injection into the national economy which may result in an overheating of the economy if domestic production is already at full capacity.

(b)   Overheating will tend to reflect itself in upward pressure on prices as total demand for goods exceeds total supply.

(c)   Surpluses are likely to put upward pressure on the exchange rate which will push up the price of exported goods in foreign countries. This gives rise to the possibility that the surplus will decline.

5.1 Goods

The visible balance is sometimes referred to as the 'balance of trade', the difference between the value of exported goods from the UK and imported goods to the UK. Prior to the growth in the importance of services, the visible balance was considered to be the major indication of the economic strength of the UK's global trade position.

Manufactured goods remain the major item of visible external trade. The balance of trade in goods has been negative in the UK for many years, having to be made up from the 'invisible sector' – see below. Note, too, that UK exports include manufacture by foreign-owned firms (such as Ford and Nissan).

5.2 Services

The invisible balance consists of services, interest, profit and dividends, and transfers. Traditionally, for example, the UK's 'invisible' balance has always shown a significant surplus, counteracting to some extent the growing visible deficit.

5.3 Interest, profits and dividends as an item of invisible trade

When a country's residents invest heavily in foreign countries, there will initially be an outflow of capital investment from that country but eventually there will be an inflow of interest, profits and dividends on those investments.

(a)   **Direct investment earnings**. These are profits of overseas branches, overseas subsidiary companies and overseas associated companies remitted to the UK. Direct investment earnings might bring income into the country (the profits of UK firms operating overseas) or cause outflows (profits of overseas firms from their investments in the UK).

(b)   **Portfolio investment earnings** (interest and dividends on stocks and shares held in securities overseas by UK residents, or held in UK securities by overseas residents).

(c)   **Interest on borrowing and lending** abroad by banks.

5.4 Transfers as an item of invisible trade

General government transfers are grants to overseas countries, subscriptions and contributions to global organisations such as the EU and other transfers by the UK government overseas or to the UK government from overseas.

6 Regional trading groups

6.1 Types of trading group

Currently, a number of **regional trading arrangements** exist, as well as global trading arrangements. These regional trading groups take three forms.

- Free trade areas
- Customs unions
- Common markets

6.1.1 Free trade areas

Members in these arrangements agree to lower barriers to trade among themselves. They enable free movement of goods and services, but not always the factors of production.

6.1.2 Customs unions

Customs unions provide the advantages of free trade areas and agree a common policy on tariff and non tariff barriers to external countries. Internally they attempt to harmonise tariffs, taxes and duties among members.

6.1.3 Economic unions/common markets

In effect the members become one for economic purposes. There is free movement of the factors of production. The EU has economic union as an aim, although not all members, including the UK, necessarily see this goal as desirable. The EU has a 'rich' market of over 300 million people and could provide a counterweight to countries such as the USA and Japan.

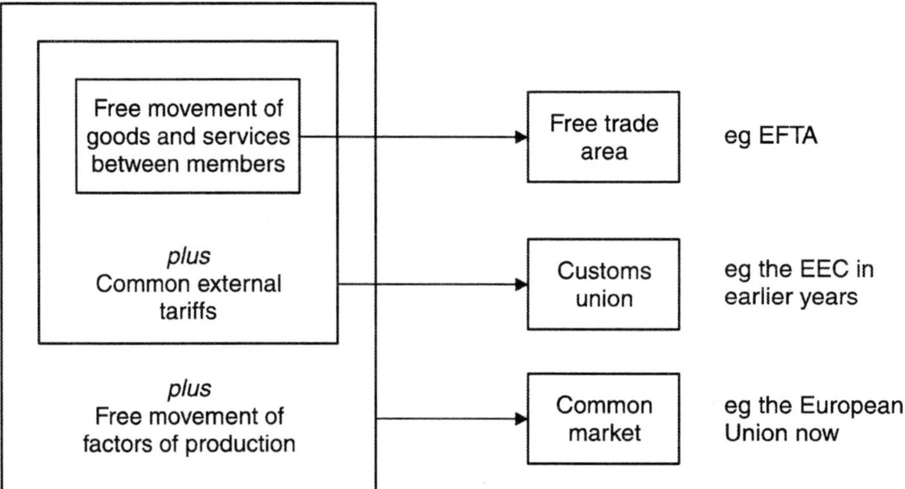

6.2 Major trading blocs

The major regional trade organisations are as follows.

North American Free Trade Agreement (NAFTA)

The United States, Canada and Mexico signed agreements on 1 January 1994 creating a trilateral trade bloc in North America. It superseded the Canada-United States Free Trade Agreement between the U.S. and Canada. In terms of combined purchasing power parity GDP of its members. In 2007 the trade bloc was identified as the largest in the world and second largest by nominal GDP comparison.

European Free Trade Association (EFTA)

Established on 3 May 1960 between European countries, the EFTA is linked to the European Union (EU). EFTA became a trade bloc-alternative for European states who were either unable to, or chose not to, join the then-European Economic Community (EEC, now EU).

Today, only Iceland, Norway, Switzerland, and Liechtenstein remain members of EFTA (of which only Norway and Switzerland are the only remaining founding members). Three of the EFTA countries are part of the European Union Internal Market through the Agreement on a European Economic Area (EEA), which took effect in 1994; the fourth, Switzerland, opted to conclude bilateral agreements with the EU. EFTA states jointly and individually form trade agreements with other countries.

European Union (EU)

The European Union (EU) is an economic and political union of 27 member states. The EU was established by the Treaty of Maastricht on 1 November 1993 to facilitate regional integration between European Communities. The EU combined generates an estimated 30% share (US$18.4 trillion in 2008) of the nominal gross world product and about 22% (US$15.2 trillion in 2008) of the PPP gross world product.

Andean Community

A trade bloc comprising the South American countries of Bolivia, Colombia, Ecuador and Peru. The trade bloc was called the Andean Pact until 1996 and came into existence with the signing of the Cartagena Agreement in 1969. Its headquarters are located in Lima, Peru.

Association of Southeast Asian Nations (ASEAN)

A geo-political and economic organisation of 10 countries located in Southeast Asia, which was formed on 8 August 1967 by Indonesia, Malaysia, the Philippines, Singapore and Thailand. Since then, membership has expanded to include Brunei, Burma (Myanmar), Cambodia, Laos, and Vietnam. Its aims include the acceleration of economic growth, social progress, cultural development among its members, the protection of the peace and stability of the region, and to provide opportunities for member countries to discuss differences peacefully.

Asean Free Trade Area (AFTA)

A trade bloc agreement by the Association of Southeast Asian Nations supporting local manufacturing in all ASEAN countries. The AFTA agreement was signed on 28 January 1992 in Singapore by six members, namely, Brunei, Indonesia, Malaysia, Philippines, Singapore and Thailand. Vietnam joined in 1995, Laos and Myanmar in 1997 and Cambodia in 1999. AFTA now comprises ten countries of ASEAN. The primary goals of AFTA seek to:

- Increase ASEAN's competitive edge as a production base in the world market through the elimination, within ASEAN, of tariffs and non-tariff barriers; and

- Attract more foreign direct investment to ASEAN.

Asia-Pacific Economic Co-operation (APEC)

A forum for facilitating economic growth, cooperation, trade and investment in the Asia-Pacific region. APEC is the only inter governmental grouping in the world operating on the basis of non-binding commitments, open dialogue and equal respect for the views of all participants. Unlike the WTO or other multilateral trade bodies, APEC has no treaty obligations required of its participants. Decisions made within APEC are reached by consensus and commitments are undertaken on a voluntary basis.

APEC has 21 members - referred to as "Member Economies" - which account for approximately 40.5% of the world's population, approximately 54.2% of world GDP and about 43.7% of world trade. APEC's 21 Member Economies are Australia; Brunei Darussalam; Canada; Chile; People's Republic of China; Hong Kong, China; Indonesia; Japan; Republic of Korea; Malaysia; Mexico; New Zealand; Papua New Guinea; Peru; The Republic of the Philippines; The Russian Federation; Singapore; Chinese Taipei; Thailand; United States of America; Vietnam.

Mercosur

A Regional Trade Agreement between Argentina, Brazil, Paraguay and Uruguay founded in 1991 by the Treaty of Asunción, which was later amended and updated by the 1994 Treaty of Ouro Preto. Its purpose is to promote free trade and the fluid movement of goods, people, and currency. The official languages are Portuguese and Spanish.

Bolivia, Chile, Colombia, Ecuador and Peru currently have associate member status. Venezuela signed a membership agreement on 17 June 2006, but before becoming a full member its entry has to be ratified by the Paraguayan parliament. The founding of the Mercosur Parliament was agreed at the December 2004 presidential summit. It should have 18 representatives from each country by 2010. Israel is currently the only non-South American free trade partner.

Southern African Development Community (SADC)

SADCC was formed in Lusaka, Zambia on April 1, 1980, following the adoption of the Lusaka Declaration – Southern Africa: Towards Economic Liberation. It was formed as a loose alliance of nine majority-ruled States in Southern Africa known as the Southern African Development Coordination Conference (SADCC), with the main aim of coordinating development projects in order to lessen economic dependence on the then apartheid South Africa.

The SADC vision is one of a common future, within a regional community that will ensure economic well-being, improvement of the standards of living and quality of life, freedom and social justice; peace and security for the peoples of Southern Africa. This shared vision is anchored on the common values and principles and the historical and cultural affinities that exist amongst the peoples of Southern Africa.

The founding Member States are: Angola, Botswana, Lesotho, Malawi, Mozambique, Swaziland, United Republic of Tanzania, Zambia and Zimbabwe.

South Asian Association for Regional Co-operation (SAARC)

The South Asian Association for Regional Cooperation (SAARC) was established when its Charter was formally adopted on December 8, 1985 by the Heads of State or Government of Bangladesh, Bhutan, India, Maldives, Nepal, Pakistan and Sri Lanka. SAARC provides a platform for the peoples of South Asia to work together in a spirit of friendship, trust and understanding. It aims to accelerate the process of economic and social development in Member States.

West African Economic and Monetary Union (UEMOA)

Members of the West African Economic and Monetary Union (also known by its French acronym, UEMOA) are Benin, Burkina Faso, Cote d'Ivoire, Guinea-Bissau, Mali, Niger, Senegal, and Togo. UEMOA member countries are working toward greater regional integration with unified external tariffs.

East African Community (EAC)

An intergovernmental organisation comprising the five east African countries Burundi, Kenya, Rwanda, Tanzania, and Uganda. The first major step in establishing the East African Federation is the customs union in East Africa signed in March 2004 which commenced on 1 January 2005. Under the terms of the treaty, Kenya, the region's largest exporter, will continue to pay duties on its goods entering the other four countries until 2010.

Global case study

The China-ASEAN Free Trade Area (CAFTA), to be established on Jan. 1, 2010, will set a good example for global cooperation, Lao Prime Minister Bouasone Bouphavanh said Tuesday.

"China and ASEAN have established strategic partnership, and it will attract much attention from the whole world," Bouasone said on the sideline of the 6th China-ASEAN Expo held from Oct. 20 to 24 in Nanning, capital of south China's Guangxi Zhuang Autonomous Region.

ASEAN, or the Association of Southeast Asian Nations, is a regional cooperation group founded in 1967. It groups Brunei, Cambodia, Indonesia, Laos, Malaysia, Myanmar, the Philippines, Singapore, Thailand and Vietnam, with a total area of 4.5 million square kilometres.

CAFTA is the first free trade area agreement signed by China, which will provide zero tariff on 90 per cent of products traded between China and ASEAN and other favourable policies on trade and investment.

"The establishment of CAFTA is both an opportunity and a challenge for China and ASEAN countries. In the future, a lot more issues concerning all aspects will need discussion," Bouasone said.

CAFTA is expected to create a combined GDP of nearly 6 trillion US dollars to become the third largest FTA in the world, only next to the North American FTA and European FTA.

"Laos is on its way to improve infrastructure and kick off construction projects, such as building railways and highways between Laos and China and other ASEAN countries, to jump into the China-ASEAN trade and investments cooperation as soon as possible," Bouasone said.

He also mentioned China, as the world's largest developing country, was a key player in the regional development and that China was of much help to ASEAN countries.

"From the Asian financial storm in 1997 to the global financial crisis starting from 2008, China has offered 10 billion US dollars worth of financial aid to ASEAN countries," Bouasone said.

http://leavefreedom.blogspot.com/2009/10/china-asean-free-trade-area-to-set.html

Accessed 29 March 2010

Discussion

Share your thoughts with your Study buddy as to the advantages and disadvantages for busines of establishing trade blocs between countries.

6.3 Regional trading agreements and the global market

Regional trading blocs only extend the benefits of free trade to their members. They may distort global trading patterns.

An idea, widely held in the 1980s, described the global business environment in the following way. This was called the 'Triad Theory'.

(a)   The industrial world was thought to be falling into three trading blocs, each led by a lead country.

- The EU (led by Germany)
- The Americas (led by the USA)
- The Far East and Pacific Rim (led by Japan)

(b)   Trading within the blocs would be relatively liberalised but there would be barriers to competition from outside. The blocs would trade with each other, but this would be more restricted. Countries would have to try to attach themselves to one of the blocs.

(c)   Non-bloc countries were not seen as terribly important. Given that some non-bloc countries, eg India and China, are likely to be some of the world's largest and fastest growing markets, suggestions to restrict trade to 'bloc' countries are short sighted. The Chinese economy, for example is growing at over 10% per annum.

The theory of trading blocs does not provide a complete analysis of world trade.

(a)   The bloc theory works better for some industries than others

(b)   The bloc theory does not really take investment flows into account

(c)   The bloc theory does not account for all trade flows. The US is Japan's most significant individual export market, although East Asia as a region has overtaken the US as a market for Japan's goods

(d)   All trading blocs have extensive economic interactions with third world countries. They will have a growing relationship with countries in the former Soviet Union and Eastern Europe, which have recently entered the global market.

A significant effect of regional trade blocs has been the rush to qualify for local status by multinational firms. This has been achieved by the multinationals setting up within one or more member states. Thus France, Germany and the UK have seen considerable inward investment from US and Japanese firms.

6.4 Benefits of regional integration

Benefit	Explanation
Trade creation	Customers have more choice of goods and services
Greater consensus	Easier to gain consensus from fewer members.
Political cooperation	Together, a few nations have significantly greater weight than if they negotiated independently

6.5 Drawbacks of regional integration

Drawbacks	Explanation
Trade diversion	Can result in reduced trade with a more efficient non-member producer and more trade with a less efficient member producer
Shifts in employment	Industries requiring unskilled labour will shift production to low-wag nations within the bloc.
Loss of national sovereignty	Successive levels of integration require nations to surrender their sovereignty.

7 International institutions and agreements

7.1 The World Trade Organisation (WTO)

The WTO was formed in 1995 as successor to the former General Agreement on Tariffs and Trade (GATT). The GATT was originally signed by 23 countries in 1947 as an attempt to promote free trade (its membership increased to 128 countries) and aimed to

- Reduce existing barriers to free trade
- Eliminate discrimination in international trade
- Prevent the growth of protectionism

GATT succeeded in reducing world tariffs substantially, after a series of 'rounds' of negotiation. The talks in the 'Uruguay round' which were concluded in December 1993 represented the culmination of seven years of work on a very ambitious programme for the liberalisation of world trade.

(a)   A major sticking point was in the area of agriculture. The European Union's Common Agricultural Policy has been seen by other GATT members as providing excessive trade protection to EU farmers.

(b)   Cultural issues were affected. The French government (despite having lured EuroDisney to the outskirts of Paris) wished to restrict imports of US movie and TV shows.

(c)   Another problem was in the textiles sector which has been subject to tariffs of over 15% and the quota levels imposed by the Multi-Fibre Arrangement.

Important facts to keep in mind about the WTO are these.

(a)   The WTO has dispute resolution powers. Aggrieved countries can take matters up with the WTO if they cannot be resolved bilaterally. For example, the EU has threatened to refer the US's Helms-Burton law, which tries to prevent other countries from trading with Cuba, to the WTO.

(b)   Membership of the WTO requires adherence to certain conditions regarding competition in the home market etc.

(c)   Membership is restricted.

7.2 United Nations Conference on Trade and Development (UNCTAD)

Dissatisfaction by underdeveloped and developing countries with GATT, due to the dominance of the rich nations, led to the formation of UNCTAD, under the aegis of the United Nations in 1964. Its influence has been modest in direct terms, but it has acted as a focus and pressure group for the less rich countries to influence other trade groups to provide special and favourable terms for underprivileged economies.

7.3 Organisation for Economic Co-operation and Development (OECD)

This is an international economic organisation of 30 countries. Most OECD members are high-income economies and are regarded as developed countries. The OECD defines itself as a forum of countries committed to democracy and the market economy, providing a setting to compare policy experiences, seek answers to common problems, identify good practices, and co-ordinate domestic and international policies.

Its mandate covers economic, environmental, and social issues. It acts by peer pressure to improve policy and implement "soft law" – non-binding instruments that can occasionally lead to binding treaties. In this work, the OECD cooperates with businesses, trade unions and other representatives of civil society. Collaboration at the OECD regarding taxation, for example, has fostered the growth of a global web of bilateral tax treaties.

7.4 The World Bank

Established in 1944, The World Bank is a vital source of financial and technical assistance to developing countries around the world. Its mission is to fight poverty for lasting results and to help people help themselves and their environment by providing resources, sharing knowledge, building capacity and forging partnerships in the public and private sectors.

They are made up of two unique development institutions owned by 186 member countries: the International Bank for Reconstruction and Development (IBRD) and the International Development Association (IDA).

Each institution plays a different but collaborative role in advancing the vision of inclusive and sustainable globalisation. The IBRD aims to reduce poverty in middle-income and creditworthy poorer countries, while the IDA focuses on the world's poorest countries.

Together, they provide low-interest loans, interest-free credits and grants to developing countries for a wide array of purposes that include investments in education, health, public administration, infrastructure, financial and private sector development, agriculture and environmental and natural resource management.

7.5 International Monetary Fund (IMF)

The International Monetary Fund (IMF) is an organisation of 186 countries, working to foster global monetary cooperation, secure financial stability, facilitate international trade, promote high employment and sustainable economic growth, and reduce poverty around the world.

With the exception of Taiwan (expelled in 1980), North Korea, Cuba (left in 1964), Andorra, Monaco, Liechtenstein, Tuvalu and Nauru, all UN member states participate directly in the IMF. Member states are represented on a 24-member Executive Board (five Executive Directors are appointed by the five members with the largest quotas, nineteen Executive Directors are elected by the remaining members), and all members appoint a Governor to the IMF's Board of Governors.

7.6 Group of Eight (G8)

The Group of Eight (G8, and formerly the G6 or Group of Six and also the G7 or Group of Seven) is a forum, created by France in 1975, for governments of six countries in the world: France, Germany, Italy, Japan, the United Kingdom, and the United States. In 1976, Canada joined the group (thus creating the G7).

In becoming the G8, the group added Russia in 1997. In addition, the European Union is represented within the G8, but cannot host or chair.

"G8" can refer to the member states or to the annual summit meeting of the G8 heads of government. The former term, G6, is now frequently applied to the six most populous countries within the European Union.

G8 ministers also meet throughout the year, such as the G7/8 finance ministers (who meet four times a year), G8 foreign ministers, or G8 environment ministers.

Each calendar year, the responsibility of hosting the G8 rotates through the member states in the following order: France, United States, United Kingdom, Russia, Germany, Japan, Italy, and Canada. The holder of the presidency sets the agenda, hosts the summit for that year, and determines which ministerial meetings will take place. Lately, both France and the United Kingdom have expressed a desire to expand the group to include five developing countries, referred to as the Outreach Five (O5) or the Plus Five: Brazil, China, India, Mexico, and South Africa. These countries have participated as guests in previous meetings, which are sometimes called G8+5.

7.7 G20

The G20 was established in 1999 to bring together major advanced and emerging economies to stabilise the global financial market.

The G20 is made up of the finance ministers and central bank governors of 19 countries plus an EU rotating Council President and European Central Bank. The countries include: Argentina, Australia, Brazil, Canada, Chine, France, Germany, India, Indonesia, Italy, Japan, Mexico, Russia, Saudi Arabic, South Africa, Republic of Korea, Turkey, United Kingdom and the United States of America.

Self-development

In order to gain an appreciation of the global marketing environment you need to put yourself into positions that give you exposure to different cultures. For example if you are considering entering the Brazilian market a good starting point is to check out any trade missions or visits that are being organised by associations in your own country in order that you can gain first hand knowledge of the market. Alternatively check out websites or media focusing on Brazil. Get to know as much as you can about the country, its people and its culture.

8 Coping with currency fluctuations: exchange rates

Global trading relationships in terms of either goods and services or in terms of capital flows result in demand for and supply of currencies in the market for foreign exchange. Most orders are quoted in hard currency (that is, a currency for which demand is persistently high relative to the supply, such as the US dollar).

These sophisticated foreign exchange markets co-ordinate the millions of interactions of demand and supply currency decisions that take place every day, leading to the determination of exchange rates between currencies.

8.1 The determination of exchange rates

Definition

An exchange rate is the price of a currency expressed in terms of another eg GBP1 = 9.06 Danish Kroner or 1 Danish Kroner = GBP0.11

The **exchange rate** between two currencies is determined by **demand** and **supply** issues.

Developing Global Marketing Strategy

(a) The exchange rate at any given time is determined by the **demand** for a currency and the available **supply**. If, for whatever reason, people want to **buy** US dollars this will exert upward pressure on the price of the dollar: people will be **selling** their own currency in exchange for dollars.

(b) People might want to buy a currency for a number of reasons.

    (i) **Trade flows**. They may need a currency to pay for goods and services. (For example, many BPP LM books are invoiced in British Pounds.)

    (ii) **Physical investment**. A Swedish organisation building, say, a hotel in Sri Lanka will need to pay its workforce in rupees.

    (iii) **Capital movements**. People wanting to invest in, say, Japanese companies listed on the Tokyo stock exchange, will need to acquire Japanese Yen to do so. Similarly people may be attracted because investments in certain currencies offer a higher rate of interest, in real terms (ie adjusting for inflation). If this increases demand, the exchange rate will rise.

(c) Obviously, trade, physical investment and capital movements are themselves influenced by a number of factors.

    (i) Economic growth.

    (ii) Importance and strength of the economy.

    (iii) Government policy: the government may have certain objectives for the exchange rate. A low exchange rate increases exports because it makes exported goods cheaper for overseas customers. The interest rates affect exchange rates.

    (iv) **The rate of inflation**. A country with high inflation will have a depreciating exchange rate. This is because if rising domestic prices are translated into other currencies, goods and services will become impossible to export.

8.2 Purchasing power parity (PPP) theory: exchange rates and inflation

Is sterling, at any moment in time overvalued, undervalued, or about right? In one sense, the answer is always 'just about right': for example the value of your nation's currency is worth whatever people are willing to pay for it in exchange for dollars, yen, euro, or whatever.

However, according to another view, the value of currency should, in the long run, tend to move towards the level of purchasing power parity (PPP), in other words the rate which equalises the domestic purchasing power of the two currencies. Thus, under PPP the exchange rate for currency at any moment will be appropriate when it is possible to buy the same basket of goods and services in various countries for the same amount of money when that money is converted at the prevailing exchange rates, adjusting of course for distribution costs, differences in consumption patterns.

While PPP theory has been found to be inadequate in explaining movements in exchange rates in the short term, it is more likely to have some validity in the long run, at least in so far as it provides a rough benchmark for interpreting movements of market exchange rates.

8.3 Government policy and the exchange rate

Exchange rate volatility is a considerable worry in economic management, especially as the flows of foreign exchange are so large. The level of the exchange rate has social consequences. The following aspects of government policy typically affect the exchange rate.

- **Interest rates**: most governments determine minimum lending rates
- **Credit controls**: some governments restrict the amount of borrowing directly
- **Inflation**: government policy influences inflation
- **Government borrowing**
- **Exchange controls**. Some governments regulate foreign exchange transactions

8.4 Fixed exchange rates

Some countries maintain a permanent fixed value of the exchange rate.

8.4.1 Advantages of a fixed rate system

The advantages of a fixed rate system are as follows.

(a)   It removes uncertainty about exchange rates and therefore about the value of goods and services. It makes global trade easier because exporters/importers can agree risk free prices.

(b)   Similarly, sellers can give buyers credit without fear of adverse movements in the exchange rate.

(c)   A fixed exchange rate system also imposes a discipline on countries to pursue responsible economic policies. In particular, there must be broadly consistent policies about inflation and economic growth.

8.4.2 Disadvantages of a fixed rate system

In order to maintain their currency at a fixed rate, the authorities must have very large quantities of foreign currency reserves, and their own currency, to intervene in the exchange market to meet an excess supply or demand for their currency.

Consequently, a government has to take other measures controlling economic activity to influence the exchange rate.

(a)   Cut imports (and boost exports) by deflationary policies to cut demand.

(b)   An inflationary policy to correct a balance of payments surplus might also be politically unacceptable to the country concerned, because inflation has social costs. As the real value of debt falls real wealth passes from lenders to borrowers and people with fixed incomes become gradually poorer.

Fixed exchange rates are often too rigid. There will occasionally be structural changes in demand and supply conditions which change the entire balance of payments prospects for a country. Different countries might have differing economic and political objectives, and so global co-operation is needed.

8.5 Floating exchange rates

Floating exchange rates are at the opposite end of the spectrum to fixed rates. At this extreme, exchange rates are left entirely to the free play of demand and supply forces.

8.5.1 Advantages of a floating exchange rate system

The advantages of free floating rates are as follows.

(a)   Government is not required to undertake difficult and unpopular deflationary or inflationary measures at home to support the exchange rate.

(b)   Uncertainty about the future value of a currency in a floating rate system can be reduced by the use of a forward exchange market, in which contracts can be made to acquire or sell currency on a future date at a specified rate of exchange.

8.5.2 Disadvantages of a floating exchange rate system

The disadvantages of a free floating system are as follows.

(a)   It is by no means certain that market forces will value the exchange rate of a currency at an 'appropriate' level, because speculation might undervalue or over value the currency.

(b)   In practice, governments cannot allow free floating, even if they wanted to. Even if they do not intervene in the foreign exchange markets, their policies on interest rates, control of the money

supply and inflation, demand management, exchange controls, import controls and employment will all have repercussions on global trade, capital flows and the exchange rate.

(c)   A free-floating exchange rate might disrupt the government's domestic economic policies. For example, if the exchange rate depreciates, import costs will rise and, in a high-importing country such as the UK, inflation will also increase.

8.6 Managed floating exchange rates

In practice, governments seek to combine the advantages of exchange rate stability with flexibility and to avoid the disadvantages of both rigidly fixed exchange rates and free floating. Managed (or dirty) floating refers to a system whereby exchange rates are allowed to float, but from time to time the authorities will intervene in the foreign exchange market.

Buying and selling in this way would be intended to influence the exchange rate of the domestic currency. Both large scale interventions (co-ordinated buying or selling by many central banks) and small scale interventions have been applied.

8.7 The costs of currency fluctuations for businesses

The cost of imports to the buyer or the value of exports to the seller might be increased or reduced by movements in foreign exchange rates. For example, if a UK importer buys goods from a US supplier for $15,000 when the exchange rate between the US dollar and £ sterling is $1.60 to £1, the importer would expect to pay £9,375 in sterling for the goods. However, if by the time the date of payment arrives, the rate of exchange is $1.50 to £1 (ie sterling has fallen in value against the US dollar) the cost to the importer would be £10,000, or £625 more than originally anticipated. The US exporter would still receive $15,000, and would not be affected by the movement.

The organisation paying in a foreign currency or earning revenue in a foreign currency therefore has a potential exchange risk from adverse movements in foreign exchange rates. Equally, if the rate of sterling against the dollar had improved over the same time span, the UK importer would have benefited.

8.8 Reducing risk in exchange

There are ways of reducing or eliminating foreign exchange exposure.

8.8.1 Bonus

Banks play a central role in the foreign exchange market, because they sell foreign currency and buy foreign currency, which is of course an essential element in global trade dealings. Banks can help traders to eliminate or minimise their foreign exchange exposure, in a number of different ways. Hedging is a general term to describe actions to minimise financial risk.

(a)   **Forward exchange contracts**. A bank will agree to sell or buy a given quantity of foreign currency at a fixed rate of exchange for delivery at a future date, some months, or even some years, ahead. Exporters and importers can therefore eliminate their foreign exchange exposure by fixing now with their bank the rate of exchange for future foreign currency revenue or payments.

(b)   **'Pure' foreign currency options**. An alternative to foreign exchange contracts, which are provided for banks by customers who want them, are currency options. These give their holder the right but not the obligation to buy or sell in the future a given quantity of a foreign currency at a fixed rate of exchange, either on or before the expiry of the option period.

8.8.2 Matching receipts and payments

When a company incurs expenditures and earns income in the same foreign currency, it can use the income to pay for the expenditures. This is the process of matching receipts and payments. Since the company will be setting off foreign currency receipts against foreign currency payments, it does not matter

whether the currency strengthens or weakens against the company's 'domestic' currency because there will be no purchase or sale of the currency.

8.8.3 Borrowing

Borrowing in a foreign currency in which payments will eventually be received, and using the payments to repay the loan.

What are the potential risk and opportunities associated with (a) fixed and (b) floating exchange rates for your organisation?

Chapter roundup

1 Appreciate the importance of the political and legal environment

- International markets offer firms opportunities to trade, to acquire resources and investment, but also offer threats in that other companies from overseas can do the same.

- We can analyse environmental factors using the mnemonic SLEPT (social, legal, economic, political, technological).

- Political factors offer extra risks to the exporter, in terms of the stability of the country, the attitude of the government to trade generally and to the company in particular.

- Legal factors in individual markets include product regulations and control over the marketing mix generally. Human resource usage is also determined by legal factors.

2 Explore the impact of protectionism in global trade

- Although trade has benefits, many countries have sought to limit its effects in order to protect local producers. In the long term this serves to harm economic welfare as resources are not allocated where they are most productive.

- Regional trading blocs promote trade between countries in groups. The environment of world trade is becoming freer, globally, with the influence of the World Trade Organisation.

3 Understand the importance of economic structure and development

- The level of a country economic development includes: infrastructure; education; ownership of durables and GDP.

- Market size will be dependent on the economic worth of a consumer market.

4 Explore the concept of the balance of payments

- Economic factors include the overall level of growth and stage of development, frequently measured by GDP. The balance of payments is an influence on government policy.

5 Understand the role of regional trading groups

- Three forms of trading groups include free trade areas, customs unions and common markets.

6   Understand the nature of international institutions and agreements

- Organisations such as the WTO, IMF, G8 and OECD exist to facilitate economic trade between countries, manage global economies and encourage global collaboration.

7   Appreciate the impact of currency fluctuations and exchange rates

- Exchange rates are the price of a currency expressed in terms of another.
- Demand and supply determines exchange rate (in floating systems).
- Exchange rate policy is determined by government.
- A spectrum of exchange rate management is possible ranging from fixed to floating.

1   The answer to this activity will depend largely on your organisation and the countries you are considering. You can treat or tolerate risks and it is for you to recommend and decide how you can counter the potential risks associated with each country.

2   It is important to be aware of the impact of protectionism policy on the ability for your organisation to succeed in certain countries. You are advised to gain an understanding of government trade policy for the country you are considering entering.

3   Be aware of the differences between the two types of exchange rates and consider the impact on your organisation's ability to trade effectively. You will need to consider this in line with the financial strategy of your organisation and should therefore work closely with colleagues responsible for the financial aspect of the organisation in order to develop a global marketing strategy that considers the financial implications.

Hollensen, S. (2007). *Global Marketing*. 4th edition. London: Prentice Hall.

Hennessey, J. (2002). *Global Marketing Strategies.* London: Economist Intelligence Unit.

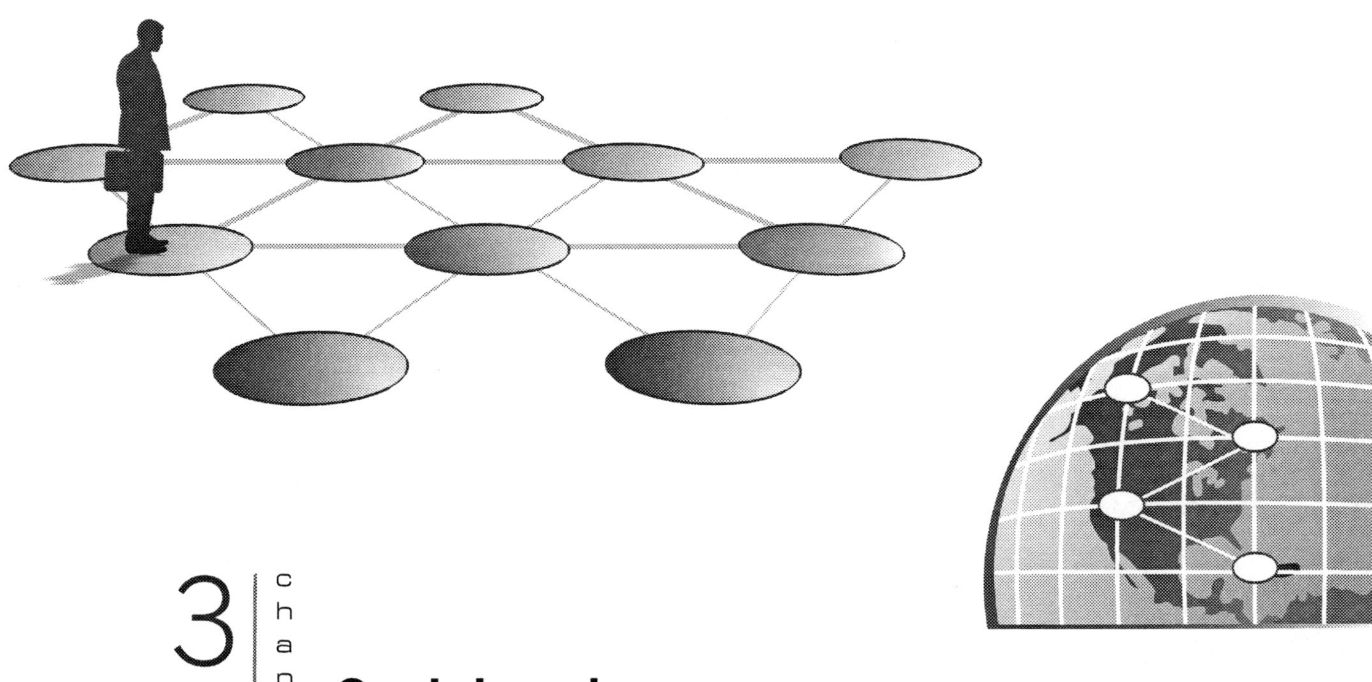

3 | chapter

Social and cultural factors in global marketing

Social and cultural factors influence all aspects of consumer and buyer behaviour. The variation between cultural and social factors in different parts of the world can be central to the development and implementation of developing successful global marketing strategies.

In this chapter we highlight the importance of culture on global marketing strategy, the impact of culture on buyer behaviour and its implications on the development of global marketing strategies. We also examine the role of culture within different aspects of business-to-business organisations and the impact of culture on these markets.

After completing this section you will be able to understand the cross cultural complexities of buyer behaviour in different markets and be able to use recognised methods to undertake cross-cultural analysis of specified global markets. We will examine issues of managing and doing business across cultures, including ethical issues in global marketing, in later sessions.

Contents

Chapter learning outcomes

By the end of this chapter you will be able to:

- Explain the components associated with social and cultural factors

- Conduct a cross-cultural analysis

- Appreciate the cultural issues associated with segmentation of global markets

- Identify the relevance of cultural factors on buyer behaviour

1 Social and cultural factors

1.1 Setting the scene

> **Definition**
>
> A **culture** is 'a set of beliefs or standards, shared by a group of people, which helps the individual decide what is, what can be, how to feel, what to do, and how to go about doing it' (Goodenough, cited by Usunier).

Culture is one of the most sensitive areas in global marketing. Culture refers to ways of feeling, behaving and '*seeing*' the world's people's basic beliefs and assumptions about who they are, what is important in life, and how they feel about themselves and their fellow humans. In marketing terms, culture has a major effect upon buyer behaviour.

Culture is particularly important for the promotion and the product elements of the mix. It is also relevant to global marketing management, in that doing business overseas requires the manager successfully to negotiate all the issues that exist behind outward business behaviour.

Of course many organisations do negotiate these problems successfully. In some cases, this is because the product is not culturally specific, but in other cases, they ensure their managers are sensitive to cultural difference. We will cover this aspect in a later chapter.

Culture is a term used by sociologists and anthropologists to encompass the total of the learned **beliefs, values, customs, artefacts** and **rituals** of a society or group.

It is clear that an individual can participate in several different cultures.

(a)   **The 'national' culture** – if such a thing exists.

(b)   **Sub-cultures** (such as 'youth culture') can be important in the early stages of product introduction. 'Computer geeks', for example, were among the first to be enthusiastic about the Internet.

(c)   **Corporate cultures** – the organisation has its own vision and values which influence the attitudes and behaviours of its people and influence its strategy and business model.

The layers or levels of a culture can be seen in the following diagram.

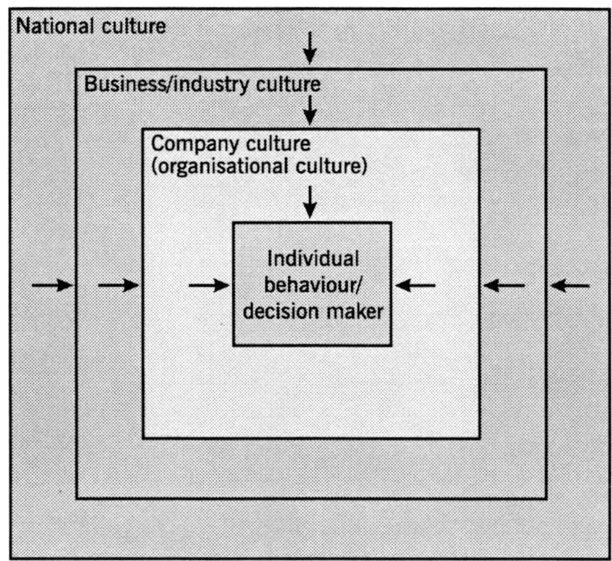

1.2 Elements of culture

'Culture' embraces the following aspects of social life.

- **Beliefs and values**. Beliefs are perceived states of knowing or cognition, on the basis of objective and subjective information. Values are the comparatively few **key beliefs**.

    - Relatively **enduring**
    - Relatively **general** – not tied to specific objects
    - Fairly widely accepted as a guide to **culturally appropriate behaviour**

- **Customs**. Customs are modes of behaviour which represent culturally approved ways of responding to given situations. There are various types of social behavioural norms. Keith Williams (1981) identifies four, on a continuum ranging from lightly to rigidly enforced patterns of behaviour, corresponding to their seriousness as a threat to social survival, if violated.

    - **Folkways** – appropriate patterns of behaviour, violation of which are noticeable

    - **Conventions** – accustomed or 'ingrained' standards of behaviour

    - **Mores** – significant social norms including moral imperatives and taboos

    - **Laws** – formal recognition of the mores considered necessary in the interests of the society as a whole, with imposed sanctions for violation

- **Artefacts**. Culture embraces all the physical 'tools' designed by human beings for their physical and psychological well-being: works of art, technology, products, buildings etc.

- **Rituals.** A ritual is a type of activity which takes on symbolic meaning, consisting of a fixed sequence of behaviour repeated over time. Ritualised behaviour tends to be public, elaborate, formal and ceremonial – like religious services, marriage ceremonies, court procedures, even sporting events.

1.3 The development of attitudes and beliefs

We acquire our feelings about the worth of goods and services through a process of cultural and social development.

In international marketing these differences in values may be quite marked, whereas in domestic markets they may be so subtle as to be unnoticeable. For the international marketer differences in attitudes and beliefs may affect the following.

- Attitude towards **ownership** of an item
- **Strength** and **direction** of attitude
- **Reason** for desire/antipathy
- **Perception** of appropriate design/style
- **Meaning** of colours, symbols and words

1.4 Language and symbols

Another very important element of culture, which makes the learning and sharing of culture possible, is language. Without a shared language and symbolism – verbal and non-verbal – there would be no shared meaning.

Verbal language is an important means of communication. The actual words spoken and the ways in which the words are spoken provide clues to the receiver about the type of person who is speaking.

Symbols are an important aspect of language and culture. Each symbol may carry a number of different meanings and associations for different people. For example, in Western cultures, the colour white symbolises purity, but in some Eastern cultures it symbolises death. Logos, trademarks and brands all have symbolic uses.

Translating words and symbols can be risky. Inadequate translation can result in clumsy errors. For example:

- The Coors beer slogan "Turn it loose" was translated into Spanish as "Suffer from diarrhoea."

- Clairol introduced its "Mist Stick" curling iron in Germany to later find out that "mist" is German slang for manure.

- Scandinavian vacuum manufacturer Electrolux used the following slogan in an American campaign: "Nothing sucks like an Electrolux".

- The Salem cigarette tag line "Feeling Free" was roughly translated into Japanese to mean "feeling so refreshed that your mind seems to be free and empty".

- An American T-shirt manufacturer printed shirts for the Latin market commemorating the Pope's visit to Miami. Instead of reading "I saw the Pope" (el Papa), the shirts read "I saw the potato" (la papa).

- The American baby food company Gerber, usually sells its jars of baby food with a picture of the Gerber Baby on the label. In certain areas of the Middle East it is customary to put pictures of the contents on the label, largely due to high illiteracy rates among women. When Gerber put the Baby on the label, people thought that it was actually made from babies.

- When McDonald's first opened in Japan the character Ronald McDonald was introduced, with his usual bright red hair and white painted face. The colour and makeup combination was representative of a death omen in Japanese culture.

The choice of language on the web is also problematic for organisations operating across many borders. Should it be multi-lingual or should it just be in one language? One preferred solution is to build a website with multilingual capabilities, localised to the language and cultural sensitivities of the market. Google for example undertakes this approach as an integral element of its business model.

Activity 1

Collect examples of how brands present themselves around the world to appeal to different cultures. There are many examples on www.youtube.com that can be downloaded. Share these with the other students in your cohort.

1.5 The transfer of cultural meaning

Some consumer researchers talk about the 'transfer of cultural meaning' at different stages of the marketing process. The 'culturally constituted world' produces products which are invested with cultural meaning or significance by advertising and 'fashion'. Those consumer goods, symbolic of cultural values, are then sold to individual consumers, who thus absorb the cultural values.

Meanwhile, by using the products in rituals of possession, exchange, grooming or whatever, the consumer adds further cultural meaning to the goods.

We have already noted that each culture establishes its own norms. Often these are derived from religious observance. Thus alcoholic drink, beef or pork, unclad females, men doing housework and so on may all be taboo in certain cultures.

Tips from the experts

Neil Stevenson (module Advisor). Some cultures are aspirational. Coca Cola appeal in many countries because of their association with the US brand. Country of origin can be an important factor in understanding the appeal of a brand/product/service eg Cuban rum, French wine and German engineering.

Global case study

Cultural differences and especially language differences have a significant impact on the way a product may be used in a market, its brand name and the advertising campaign. Coca-Cola had to withdraw their two-litre bottle from Spain when they found that Spaniards did not own fridges with sufficiently large compartments. Johnson's floor wax was doomed to failure in Japan as it made the wooden floors very slippery and Johnson's failed to take into account the custom of not wearing shoes inside the home. Initially, Coca-Cola had enormous problems in China as Coca-Cola sounded like 'Kooke Koula' which translates into 'A thirsty mouthful of candle wax'. They managed to find a new pronunciation 'Kee Kou Keele' which means 'joyful tastes and happiness'. Other companies who have experienced problems are General Motors which experienced difficulties with its brand name 'Nova' in Spain ('no va' in Spanish means 'no go') and McDonald's whose character Ronald McDonald failed in Japan because his white face was seen as a death mask.

On the other hand, there are visible trends those social and cultural differences are becoming less of a barrier. This has led to the emergence of a number of world brands such as Microsoft, Intel, Coca-Cola, McDonald's, Nike etc, all competing in global markets that transcends national and political boundaries. However, there are a number of cultural paradoxes which exist. For example, in Asia, the Middle East, Africa and Latin America there is evidence both for the westernisation of tastes and the assertion of ethnic, religious and cultural differences. These differences do not necessarily constitute unbridgeable cultural chasms in all sectors of a society. Instead there are trends towards similarities both in cultures and outlooks of consumers. There are more than 600000 Avon ladies now in China and a growing number of them in Eastern Europe, Brazil and the Amazon.

2 Cross-cultural analysis

2.1 Overview

There are approximately 200 different countries, each of which contains many different cultures with a myriad of beliefs, norms and taboos. It is helpful for us to have a framework showing the major categories.

Taking the example of religion, this can affect market entry (religious conflict) marketing organisation (days of prayer) goods (type of clothes worn, types of food eaten) promotion (images shown), methods of doing business (eg Islamic banks) and so forth.

Global marketers must decide the relevant cultural segments/groupings for analysis and then use appropriate frameworks to enable comparisons to made and contrasts and similarities to be drawn across cultural groupings.

2.2 National cultures

Many cultures can be identified with a nation, but often this does not apply.

- The UK (for example) is relatively small, with one language, English, predominantly spoken (despite Welsh also being an important language).

- A large country like India is home to many language groups, and while Hinduism is shared across many of these language groups, there is also large Muslim community

- Some countries combine two more or less equal linguistic cultural groups. For example Switzerland is a nation state that is 'explicitly multi-cultural'.

Many of the world's nation states, with defined borders and central institutions, are relatively recent creations, resulting not from ethnicity or cultural homogeneity but from colonial administration. This is particularly true of Africa.

2.3 High context and low context cultures

A distinction needs to be drawn between **high context** and **low context** cultures. All communications have a **context,** which can influence the message.

- **Where** the communication takes place
- The **people** involved
- The **content** of the conversation (eg work, negotiation)

Context will often shape communication and can indicate how the content of a message should be interpreted.

(a) **Low context** cultures are ones in which context is unimportant to the meaning of a message. Like railway timetables, words should mean exactly what they say. The meaning is explicit.

(b) In **high context** cultures, messages must be interpreted from the context. In a high context culture, communication might be impossible if you do not know the person you are with.

Usunier (2000) identifies the following groups.

- Low-context and explicit messages: Swiss, Germans, Scandinavians, North Americans
- Medium-context: English, French, Italian, Spanish
- High context: Latin Americans, Arabs, Japanese

The following diagram looks at a variety of nations in terms of the contextual differences.

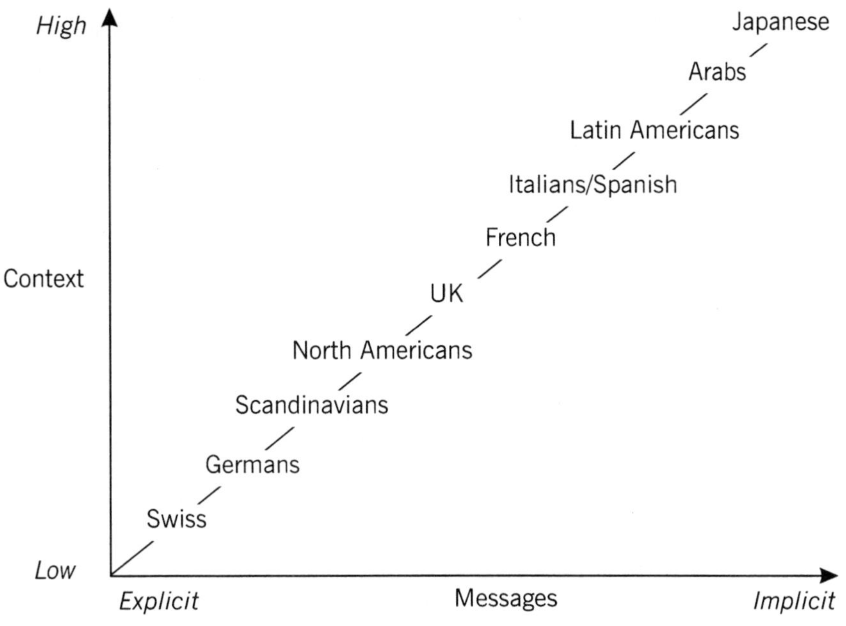

Adapted from Usunier (2000).

2.4 Hofstede's model of cultural dimensions

A model was developed in 1980 by Professor Geert Hofstede in order to explain national differences by identifying 'key dimensions' which represent the essential 'programmes' forming a common culture in the value systems of all countries. Each country is represented on a scale for each dimension so as to explain and understand values, attitudes and behaviour.

In particular, Hofstede pointed out that countries differ on the following dimensions.

(a)   **Power-distance**. This dimension measures how far superiors are expected to exercise power. In a high power-distance culture, the boss decides and people do not question.

(b)   **Uncertainty-avoidance**. Some cultures prefer clarity and order, whereas others are prepared to accept novelty. This affects the willingness of people to **change** rules, rather than simply obey them.

(c)   **Individualism/collectivism**. In some countries individual achievement is what matters. A collectivist culture (eg people are supported – and controlled – by extended families) puts the interests of the group first.

(d)   **'Masculinity'/'Femininity'**. In 'masculine' cultures, gender roles are clearly differentiated. In 'feminine' ones they are not. 'Masculine' cultures place greater emphasis on possessions, status, and display as opposed to quality of life, the environment etc.

Hofstede grouped countries into eight 'clusters'.

Group		Power-distance	Uncertainty avoidance	Individual-ism	'Masculinity'
I	'More developed Latin' (eg Belgium, France, Argentina, Brazil, Spain)	High	High	Medium to high	Medium
II	'Less developed Latin' (eg Portugal, Mexico, Peru)	High	High	Low	Whole range
III	'More developed Asian' (eg Japan)	Medium	High	Medium	High

Group		Power- distance	Uncertainty avoidance	Individual- ism	'Masculinity'
IV	'Less developed Asian' (eg India, Taiwan, Thailand)	High	Low to medium	Low	Medium
V	Near Eastern (eg Greece, Iran, Turkey)	High	High	Low	Medium
VI	'Germanic' (eg Germany)	Low	Medium to high	Medium	Medium to high
VII	Anglo (eg UK, US, Australia)	Low to medium	Low to medium	High	High
VIII	Nordic (eg Scandinavia, the Netherlands)	Low	Low to medium	Medium to high	Low

There are dangers in using these models. In the management of individual businesses, other factors may be more important.

(a) **Type of industry**: people working in information technology from two countries might have more in common with each other than they might with people working in a different industry.

(b) **Size of company**. Some people may be accustomed to working in a bureaucracy.

2.5 Hofstede's cultural dimensions versus rate of product adoption

Singh (2006) suggests that particular dimensions of culture are critical in determining whether consumers of certain cultures are likely to easily adopt new products or services.

Singh found that cultures characterised by a small power distance, weak uncertainty avoidance and masculinity are more likely to be innovative and accept new product ideas than cultures where there is a large power distance, strong uncertainty avoidance and more feminine traits.

He also identified that more individualistic cultures are likely to be more impersonal channels while collectivist cultures are swayed by more interpersonal communications.

Discussion

With reference to Hofstede's model compare and contrast your country culture with your Study buddy and post your findings in the discussion forum. Consider the impact of culture on the way you are studying for this qualification. You can also visit http://www.geert-hofstede.com/

3   Culture and global marketing

Cultural segmentation must be considered particularly carefully, therefore, in an international (or **cross-cultural**) context the marketer needs to understand the beliefs, values, customs and rituals of the **countries in which a product is being marketed in order to alter or reformulate the product to appeal to local needs and tastes, and reformulate the promotional message to be intelligible and attractive to other cultures.**

- Nestlé, the Swiss coffee maker, sells different strengths and styles of coffee in Europe than in America. Haagen-Dazs developed green tea ice-cream for the Japanese market.

- Board games such as Monopoly are sold worldwide – with nationally-relevant street/area names, money and tokens.

- Legal and regulatory provisions with regard to advertising vary from country to country: the showing of cigarettes in ads, for example, the use of comparative advertising, or the use of children in advertising.

- Products are positioned as exotic imports with the specific appeal of their country of origin (Australian lager, Italian pizza, French mineral water) – while in the domestic market they are sold on familiarity and cultural loyalty. There are also examples where the home country does not buy the product eg Fosters, Blue Nun etc

There are two ways of looking at cross-cultural marketing:

(a) **Localised marketing strategy** stresses the diversity and uniqueness of consumers in different national cultures.

(b) **Global marketing strategy** stresses the similarity and shared nature of consumers worldwide. We refer to this as 'globalisation'.

The marketer will be interested in the extent of the cultural differences between two countries with regard to the following:

		Potential problems
(a)	**Language**	A promotional theme may not be intelligible – or properly translatable – whether in words or symbols.
(b)	**Needs and wants**	The benefits sought from a product in one country may be different in another.
(c)	**Consumption patterns**	One country may not use a product as much as another (affecting product viability), or may use it in very different ways (affecting product positioning).
(d)	**Market segments**	One country may have different demographic, geographical, socio-cultural and psychological groupings to another.
(e)	**Socio economic**	Consumers in one country may have different disposable income factors and/or decision-making roles from those in another.
(f)	**Marketing conditions**	Differences in retail, distribution and communication systems, promotional regulation/legislation, trade restrictions etc may affect the potential for research, promotion and distribution in other countries.

A **failure to understand cultural differences** may cause problems with each of the elements of the marketing mix:

(a) **The product**. Nestlé coffee, Camay soap and a host of similar products are marketed internationally – with different names, flavours, aromas, and other characteristics. Coffee is preferred very strong, dark and in ground form in Continental Europe – but weaker, milder and instant in the USA. Colour is another interesting element in product and packaging: it means different things to different cultures. Blue is said to represent warmth in Holland, coldness in Sweden, purity in India and death in Iran. Product benefits also vary in appeal: Pepsodent apparently tried to sell their toothpaste using the slogan 'You'll wonder where the yellow went!' in South-East Asia – where chewing betel nuts is considered an elite habit (giving status-enhancing brown teeth). Products themselves are more or less popular in different countries. Malt based soft drinks are highly popular in Ghana for example but are not necessarily drunk in this format outside of Africa.

(b) **Promotion**. Customs, symbols and language do not always travel well. 'Come alive with Pepsi' was translated as 'Pepsi brings your dead ancestors back to life' in one Far Eastern country.

(c) **Price**. Large package sizes may not be marketable in countries where average income is low, because of the cash outlay required.

(d)   **Place**. Distribution tastes may vary. Supermarkets are very popular in some countries, while others prefer intimate personal stores for groceries and other foodstuffs. Newly-opening markets such as Eastern Europe exhibit poor distribution systems and very low salesperson effort and productivity – compared to Japan, say, which rates very highly. Even outwardly similar cultures for example in terms of food purchasing habits can in practice be dissimilar. A UK premium breakfast cereal manufacturer had multitude of packaging sizes because of differences in their export markets. Malta for example had many smaller store formats and top-up shopping was a common behaviour meaning smaller size packs were required. Conversely Canadians shopped very infrequently and therefore demanded very large packs which were distributed by superstores.

(e)   **People**. There are differing attitudes to personal service in some countries. Clearly different countries have different service cultures.

(f)   **Processes**. A good example of the cultural influence on processes in some industries is the religious requirements regarding slaughtering and cooking of halal and kosher food.

(g)   **Physical evidence**. In service industries such as hotels, physical evidence is of the 'essence'. Different people have different expectations as to requirements for comfort.

International business activities introduce new products, services and ways of doing business. Some cultures will resist this if no attempt is made to understand the cross-cultural dimensions of global marketing. Consideration has to be given to understanding the impact of the firm's activities.

Global marketers need to analyse the different elements of culture in order to develop an effective strategy. They need to think about the following:

- Customer needs and motivations
- Buyer behaviour
- Cultural values relevant to the product
- Decision making in the target market
- Appropriate promotion methods
- Appropriate distribution channels

The growth of the **service sector**, and the marketing of services, is affected by culture. Services such as travel, culture and health and fitness are becoming status symbols in many European consumer groups. Culture probably affects service products more than goods, and so the marketing of such services has to be aware of regional differences.

3.1 Cultural convergence and divergence

For certain product groups, where social exposure is important, and to the social classes where status is also important, the acquisition and adoption of an international lifestyle has led to international products such as Walkman, Coke etc. Two groups seem to be most affected.

- **Young people**, who have an above average need for social acceptance
- **International travellers**, who are exposed to multicultural values

This **convergence** process has led to the idea of global products that have an appeal worldwide with little or no modification.

3.2 Self reference

Another problem is self reference, in other words the tendency to interpret something according to one's own values and experience, rather than on its own terms. It is pointless getting upset about practices which are 'the norms' in the overseas market.

To avoid misinterpreting other people in the light of one's own assumptions, some firms analyse their products and promotional messages specifically for their cultural connotations, and how the product relates to them.

- Motivation? What needs are satisfied?
- What are the characteristic patterns of behaviour?

- What values are relevant to the product?
- How are decisions taken?
- Promotion messages: taboos, language etc.
- Institution – what types of retailing are acceptable for a product?

The marketer has to **reconcile the respective cultures** of the seller and the buyer. Trompenaars and

Trompenaars and Woolliams (2003) highlight seven dimensions to the international marketer's dilemma in addressing this issue. They are presented in the table below.

Cultural dimension	Question	Example
Universal or particular	'Do we follow a single global approach or particularise to each market'?	**Transnational specialisation.** Coca-Cola has different recipes for different cultures, depending on the level of sweetness required by local palates.
Individualism or **communitarianism**	'Is marketing concerned with satisfying individual needs, or in creating a trend that is adopted by the group?'	The relationship should be seen as circular. Microsoft Windows and its 'Office' products offer the benefits of a group approach through share/exchange of documents, but systems can be configured to individual preferences.
Specific or **diffuse**	'Is the customer a "punter" or a series of relationships?'	The marketing team needs to identify how the relationship can be deepened and provide a more **personal service.** American Airlines emphasises shareholder value, while British Airways (its One World partner) emphasises service, hot breakfasts and champagne. Who is right?
Neutral or **affective**	'What part does emotion play?'	**Adapt your message to the culture.** Michael Porter (*Competitive Advantage of Nations*, 1998) says that the American way of displaying the qualities/features of your product may be interpreted by Germans as bragging, so marketers need to be more subtle.
Achievement or **ascription**	'Do customers want a functional product that achieves its purpose, or are they buying status?'	Marketers need to **care about the people they serve**, and ascribe status to them. BUPA reconciles the need to achieve business growth with the provision of health care.
Internal and **external control**	'Do we have an inner drive, or do we adapt to external events?'	**Need to reconcile 'pull' and 'push'.** Philips has great technology and marketing, but do the two connect properly? 'The push of the technology needs to help you decide what markets you want to be pulled by, and the pull of the market needs to help you know what technologies to push.'
Time	'Do we view time as sequential or synchronic?'	Do you strive to get to the market **quickly**, or do you get there **'just in time'** to meet needs? Tesco initially stole a march on the other UK supermarkets with its innovative and popular Internet home delivery service, which required considerable investment even before it could be sure of success.

Saudi Arabia is the sixth largest fragrance market and has the world's highest per capita consumption of fragrance. Big fragrance houses generally use the same promotional materials used by marketers in Europe. Usually Saudi Arabia is a high-touch culture but a Drakkar Noir advertisement toned down the sensuality for the Arab version of an advertisement for the men's perfume. The European advert featured a man's hand clutching the bottle and a woman's hand seizing his bare arm. In the Saudi version, the man's arm was clothed in a dark jacket sleeve, and the woman touched the man's hand only with her fingertip (adapted from Hollensen, 2007).

4 Cross-culture and buyer behaviour

While we have examined the effect that culture has on the marketing mix, it might be worth examining in a little more detail why this should be so, with reference to motivation and buyer behaviour.

Some writers, like Abraham Maslow (1987), believe that all motivation is the result of a desire to satisfy one of several needs. He devised a hierarchy of needs. These are outlined below. You will have encountered these before.

- Physiological
- Safety
- Love/social
- Esteem
- Self-actualisation
- Freedom of enquiry and expression
- Knowledge and understanding needs

The 'ethnocentricity' of Maslow's hierarchy has been noted. For example it is simply not true that in every culture the needs at a definite level must be satisfied in order for higher order needs to appear.

Furthermore similar kinds of needs may be satisfied by very different products and consumption types.

Also different cultures focus on different needs. For example Hinduism encourages self-actualisation and discourages the pursuit of lower level needs. In some developing cultures satellite TV may be more important than the need for food.

However, it is a useful model of the sort of needs people might have. Cultural factors in a society influence a number of the needs Maslow identifies.

- **Physiological needs** are likely to be powerful in societies where people are poor or where extended families and kinship networks place special demands on the working individual's role as **provider**.

- **Safety needs** are powerful in particular social contexts.

    - For example, the practice of lifetime employment in some Japanese companies began after World War II. Safety was promised as a 'job for life' in return for dedication to work. Economic factors are now taking their toll on this concept and the 'job for life' cannot now be guaranteed.

    - Just as importantly, perception of 'safety' and 'risk' differ. Some societies have a variety of different attitudes to 'food' and 'health' and the risk of infection, which can have an important effect on the desirability of health products.

- Satisfaction of **esteem needs**

    - The visible trappings of success
    - The respect of the group

- **Self-actualisation**, the fulfilment of personal potential may not be seen as a 'need' in some societies, especially those where the success of the 'group' and conformance to its demands is the dominant value system. 'Self-actualisation' implies an individualist philosophy not always present in many cultures.

It is worth emphasising that buyer behaviour is influenced by a number of factors, not only motivational issues. Here are some examples.

- Who takes major purchase decisions? The husband or wife? Or are they reached by group discussion, as might be the case in an extended family? Do people defer to the elders in their family? Is individualism or collectivism the cultural orientation?

- What is the overall buying power of an individual in a buying situation?

- What is the retailing structure (eg large out of town supermarkets encourage stocking up once a week)? To what extent is the population urban or suburban?

Culture is an important variable in the purchase decision.

- Culture can render certain products unacceptable in the first place (eg advertising pork or beef in countries where there are taboos against eating pigs or cows).

- Culture can also bias a person's perception of information.

Usunier (2000) cites institutions, social conventions, habits and customs as relevant to buyer behaviour. He gives the example of eating. Cultures differ in the following ways.

- The number of meals consumed in a day
- The duration of meal times
- The composition of the meal (cooking style, portion size)
- The extent to which a meal is just a 'refuelling' stop or a family/social event
- Is the food prepared with basic ingredients or is it purchased part-cooked?

The list is endless because nothing is more essential, more universal and at the same item more accurately defined by culture than eating habits. Eating habits should be considered as the whole process of purchasing food and beverages, cooking, tasting and even commenting. In many countries, commercials advertising ready-made foods (canned or dried soups, for instance) faced resistance in the traditional role of the housewife, who was supposed to prepare meals from natural ingredients for her family. As a result, advertisers were obliged to include a degree of preparation by the housewife in the copy strategy.

4.1 Social and cultural influences in business-to-business marketing

Most models of business buyer behaviour refer to the decision-making unit, in other words the group of people responsible for making the decisions whether to buy a product. We can examine some possible problems below.

(a) **Authority and delegation**. In some organisations, individuals have clearly defined areas of authority and decision-making power. However in some cultures, decisions have to be referred upwards to more senior figures, so the person doing the negotiating may not have the right to make the decision.

(b) **Clarity of authority**. In some cultures, managerial decision making is taken by consensus. The problem is to manipulate this consensus on your behalf. The way in which decisions are taken or can be overridden can be a significant problem.

(c) **The decision process**. Does the DMU judge proposals according to the 'rational model'? In other words, are a number of alternatives evaluated in the cold light of day, or do other 'political' considerations intrude? If the firm is part of a network of other firms, it might be under pressure to buy from a group company.

A factor which has an impact on organisations engaging in global marketing is the management culture.

This comprises the views about managing held by managers, their shared educational experiences, and the 'way business is done'. Obviously, this reflects wider cultural differences between countries, but national cultures can sometimes be subordinated to the corporate culture of the organisation.

A 'world leadership survey' conducted by the *Harvard Business Review* asked a variety of questions to managers in different countries. Managers in different countries do not seem to have the same priorities when it comes to business issues. When asked what they thought were the three most important factors in organisational success, these were listed as follows, in order of priority.

- Japan: product development, management, product quality
- Germany: workforce skills, problem solving, management
- USA: customer service, product quality, technology

Cultural differences might affect buying behaviour, especially during negotiations.

(a)   How do you establish the salesperson's credibility? Many cultural preconceptions can underline the difficulties of assessing which person to believe. For example, frankness is to be avoided in some cultures (if it means someone else is losing face).

(b)   Is the style of negotiation communicative or manipulative? In other words, do you want to exchange facts or manipulate the other party?

(c)   To what extent are oral agreements the basis for business, and to what extent are contracts or written agreements preferred?

We will look at these issues in greater depth in later sessions.

4.2 Government buyer behaviour

In many countries, government is the biggest buyer and will be responsible for buying a wide range of goods and services. The way government buys is influenced by the extent to which public accountability for expenditure is deemed to be important – so a cultural history of public service and accountability is critical.

The usual forms of buying procedure are the open tender and the selective tender. For a selective tender process, the firm needs to be accepted on the appropriate list. In some countries, it takes considerable persistence to get to that stage, since it may take several visits to appropriate government officials to establish a good working relationship.

In the EU, most public procurement contracts should be open to any competitor throughout the European Union – and the bidding will normally be to an open tender.

It is generally true that in dealing with governments and government departments, the significance lies, perhaps, with who you know rather than with what you know. The right political contacts are often essential.

It is also true that buying decisions may be affected as much by a lowly departmental clerk acting as a gatekeeper in a Decision Making Unit (DMU), as by a senior government official or even a politician acting as a decision maker or an influencer in the DMU. A clerk in a regional office can make life very difficult, despite support at the centre.

A clear understanding of the cultural inter-relationships in government circles, hierarchical relationships and political influence may be critical in some markets and countries.

(a)   For example, a different style is necessary for dealing with officials from high context cultures than from low context cultures.

(b)   Officials may have formal authority, but little actual power.

(c)   Firms suffer if relationships between the domestic government and the host country government deteriorate.

(d)   Different cultures have different attitudes to gifts.

(e)   Some cultures prefer a high degree of legalism, others do not.

(f)   There may be conflict between different government departments.

(g)   Governments are susceptible to pressure from powerful interest groups at home.

We will explore the impact of culture on how to conduct business with different organisations around the world and the cultural dimensions on ethical decision making in a later session.

Assessment advice

This session has focussed on developing an understanding of the components of culture and how these components impact on consumer beliefs, values, attitudes and purchasing behaviour. Culture has a significant impact on global marketing strategies of organisations, both within business-to-consumer and business-to-business markets. Consider the relevance of the models and theories and their impact on the global marketing strategy you are proposing for the company under investigation in your assignment. Ensure that you apply them appropriately in the consideration of the recommendations you are making.

4.3 Culture and management

As we have discussed, global marketers must decide to what extent adaptation to given cultural specifics need to be considered.

For example in low context cultures such as Germany, Switzerland and Austria punctuality is considered very important and highly valued within those cultures. By contrast in Latin American countries a looser interpretation of time is accepted. This illustrates the differences within cultural groups; one is neither right nor wrong, just different.

In attempting to understand another culture we inevitably interpret new cultures through our own eyes based on our understanding of our own culture.

Therefore it is particularly important to understand new markets in the same ways as buyers or potential buyers.

Lee (1996) used the term 'self reference criterion' (SRC) to characterise our unconscious reference to our own cultural values. Lee suggested a four-step approach to eliminate SRC:

1   Define the problem or goal in terms of home country culture, traits, habits and norms

2   Define the problem or goals in terms of the overseas country culture, traits, habits and norms

3   Isolate the SRC influence aspect of the problem and examine it carefully to see how it complicates the problem

4   Redefine the problem without the SRC influence and solve for the overseas market situation

4.4 The effect of culture on ethical decision making

Culture is a fundamental determinant of ethical decision-making and directly affects how an individual and an organisation perceives ethical problems and consequences.

In order to succeed in today global markets, marketers need to understand and recognise how these issues differ across cultures and influence marketing decision making.

Every culture establishes a set of moral standards for business behaviour which it refers to as its code of ethics. This set of standards influences all decisions and actions and is heavily influenced by the culture in which it takes place.

Ethical business conduct should normally exist as a level well above the minimum required by law or the "controlling legal authority".

Three ethical principles also provide a framework to help the marketer distinguish between right and wrong and what ought to be done.

1   **Utilitarian ethics**
    Does the action optimise the "common good" or benefits of all constituencies?

2   **Rights of the parties**
    Does the action respect the rights of the individuals?

3   **Justice or fairness**
    Does the action respect the canons of justice or fairness to al parties involved?

Answers to such questions will help the global marketer ascertain the degree to which decisions taken are beneficial or harmful, right or wrong and whether the consequences are ethical or socially responsible.

Global case study

In December 2005 Primark, the UK-based discount clothing chain was rated the least ethical place to buy clothes in Britain.

Primark scores just 2.5 out of 20 on an ethical index that ranks the leading clothing chains on criteria such as workers' rights and whether they do business with oppressive regimes. Mk One and Marks & Spencer were ranked second and third worst for ethics by Ethical Consumer magazine.

As Primark has grown, its record on ethical trading has come in for close scrutiny. People question how it can offer good quality fashion at low prices. Its success is based on big volumes, low mark-ups, and minimal advertising. It is a lean business which responds quickly in the marketplace, has short lines of management, good buying and excellent distribution. It believes it is firm but fair with its suppliers, and offers terms that compare favourably with those of its competitors.

It shares more than 95% of its factories with other brands. It also shares with its competitors many of the challenges confronting any retailer with a global supply chain. It says on its website it is firmly committed to improving the ethical performance of its business and that of its suppliers and their factories.

When selecting new suppliers, it requires them to go through a selection process which entails a comprehensive audit of labour standards, against its Code of Conduct by a trusted third party. Satisfactory performance, including completion of any necessary corrective action requiring immediate remediation, is a pre-requisite for working with them. They have approximately 600 first tier suppliers whom they buy from directly.

Adapted from:http://www.independent.co.uk/news/uk/this-britain/primark-is-named-as-least-ethical-clothes-shop-518600.html

http://www.primark.co.uk/Ethical

Accessed 2 April 2010

Chapter roundup

1   Explain the components associated with social and cultural factors

- Cultures are exemplified in language, religions, customs, value systems, education, law and aesthetics.

- A variety of frameworks exist to analyse culture. For example, low context cultures require messages to be clear and direct. High context cultures place more emphasis on body language, and messages are not always explicit.

- Some cultures will resist new products of services that you may wish to implement if no attempt is made to understand cross-cultural factors.

- Different types of buyer in different organisational settings display different attitudes, needs and purchasing behaviour.

- Culture is one of the most sensitive areas in global marketing.

- Culture is particularly important for the promotion and the product elements of the mix.

- Culture is a term used by sociologists and anthropologists to encompass the total of the learned beliefs, values, customs, artefacts and rituals of a society or group.

- It is clear that an individual can participate in several different cultures: the 'national' culture, sub-cultures and corporate cultures.

- Another very important element of culture, which makes the learning and sharing of culture possible, is language.

- Translating words and symbols can be risky. Inadequate translation can result in clumsy errors.

- There are approximately 200 different countries, each of which contains many different cultures and with a myriad of beliefs, norms and taboos.

- It is helpful for us to have a framework showing the major categories.

2 Conduct a cross-cultural analysis

- Global marketers must decide the relevant cultural segments/groupings for analysis

- Many cultures can be identified with a nation, but often this does not apply.

- A distinction needs to be drawn between high context and low context cultures.

- A model was developed by Professor Geert Hofstede to explain national differences

3 Appreciate the cultural issues associated with segmentation of global markets

- Singh suggests that particular dimensions of culture are critical in determining whether consumers of certain cultures are likely to easily adopt new products or services.

- Cultural segmentation must be considered particularly carefully in an international (or cross-cultural) context.

- The marketer needs to understand the beliefs, values, customs and rituals of the countries in which a product is being marketed.

- There are two ways of looking at cross-cultural marketing – localised marketing strategy and global marketing strategy.

- A failure to understand cultural differences may cause problems with each of the elements of the marketing mix.

4 Identify the relevance of cultural factors on buyer behaviour

- Culture can be an influence on buyer behaviour, but it should be noted that relatively simple models of motivation, such as Maslow's, do not always apply across cultures

- Global marketers need to analyse the different elements of culture in order to develop an effective strategy.

- The marketer has to reconcile the respective cultures of the seller and the buyer.

- Different cultures focus on different needs.

- Buyer behaviour is influenced by a number of factors, not only motivational issues.

- Culture is an important variable in the purchase decision.

- Cultural differences might affect buying behaviour, especially during negotiations.

- In many countries, government is the biggest buyer and will be responsible for buying a wide range of goods and services.

- A clear understanding of the cultural inter-relationships in government circles, hierarchical relationships and political influence may be critical in some markets and countries.

- It is important to understand new markets in the same ways as buyers or potential buyers.

- Marketers need to understand and recognise how cultural issues differ across countries and influence marketing decision making.

- Ethical business conduct should normally exist at a level well above the minimum required by law or the "controlling legal authority".

- Three ethical principles also provide a framework to help the marketer distinguish between right and wrong and what ought to be done: utilitarian ethics, rights of the parties and justice or fairness.

1   This answer will depend upon the examples that you find.

Cateora, P. and Graham, J., (2009). *International Marketing*. 14th edition. London: McGraw Hill.

Doole, I. and Lowe, R., (2008). *International Marketing Strategy Analysis, Development and Implementation*. 5th edition. London: Thomson Learning.

Hofstede, G., (1980). *Culture Consequences: International Differences in Work Related Values*. Sage Publishing.

Hollensen, S., (2007). *Global Marketing*. 4th edition. London: Prentice Hall.

Maslow, A., (1987). *Motivation and Personality*. New York: Longman.

Trompenaars, F., and Wooliams, P., (2003). *Business Across Cultures*. London: Capstone Publishing.

Usunier, J-C., (2000). *International Marketing*. London: Pearson Education.

Usunier, J-C. and Anne Lee, J., (2009). *Marketing Across Cultures* .5th Edition. London: Prentice Hall.

Williams, K., *Behavioural Aspects of Marketing*. Oxford: Butterworth-Heinemann.

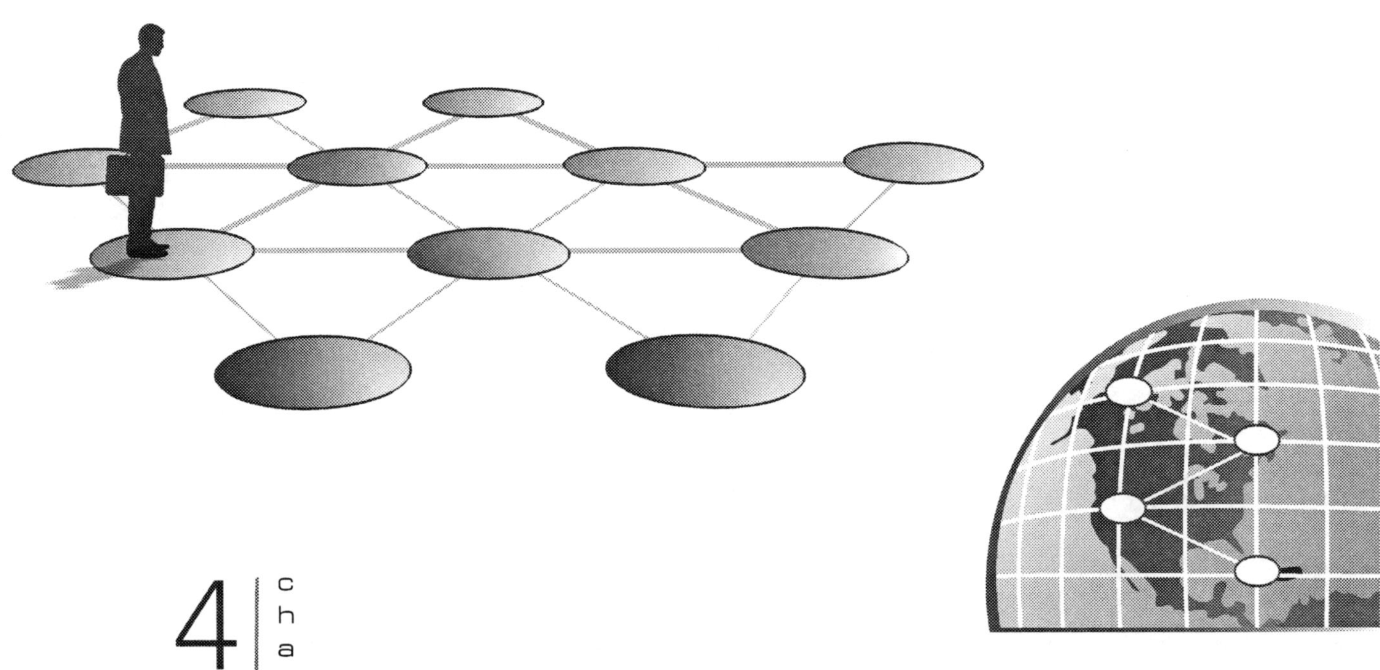

4 | chapter

Developing a global vision through marketing research

The global marketer works in a highly complex and risky environment. If an organisation is to survive and thrive in the global marketplace then it is important to search for methods that reduce the risk of making a wrong decision as far as is possible. Global marketing research (GMR) is critical to the marketing process. Lack of knowledge of global markets is one of the first major barriers that must be overcome and the successful marketer will develop an effective marketing research strategy that overcomes that barrier.

The purpose of this chapter is to examine the role of marketing research in marketing strategy and its contribution to the decision-making process. This chapter examines the role of global marketing research at a global level, the objectives, techniques, opportunities and limitations and the development of a global marketing research system within the organisation.

Contents

Chapter learning outcomes

In this chapter you will cover the following:

- Understand the role of global marketing research
- Evaluate information strategy and sources
- Understand databases and expert systems
- Appreciate the GMR process
- Understand the uses of GMR data
- Appreciate some problems in GMR
- Understand how to manage and implement the GMR process

1 Setting the scene

While the principle of marketing research remains the same for global markets as for domestic markets, global marketing research (GMR) is often different in practice from research in the domestic market, not least because of the sources of data that may be available. Not only are countries different from each other, but there are the major problems of distance and the quality of data to contend with.

The purpose of GMR is to identify suitable countries for entry, or suitable groups of countries whose populations segment in a way that can be met by a similar marketing mix. Much secondary data is poor, and the organisation is even more 'in the dark' about what is going on than they might otherwise suppose.

There is greater reliance on human as opposed to documentary sources. Even primary data has to be used with care. The organisation of the GMR function (centralised at head office or decentralised through country subsidiaries) is analogous to most organisational problems in global marketing.

Definitions

Primary data is information that is collected first-hand, generated by the original research undertaken to answer specific research questions.

Secondary data is information that has already been collected for other purposes and is readily available.

Assessment advice

Research and the underlying problems involved in researching global markets is an essential element of developing a global marketing strategy. You should collect secondary data and undertake some primary research to help you determine the direction your organisation should be taking and the country you are proposing the organisation under investigation should enter.

2 The role of Global Marketing Research (GMR)

2.1 Reasons for global marketing research

Definition

Marketing research is the objective gathering, recording and analysing of all facts about problems relating to the transfer and sales of goods and services from producer to consumer or user.

Developing Global Marketing Strategy

The vast majority of organisations, particularly SMEs, place little priority on GMR. Where any intelligence is gathered it is usually through intermediaries rather than gleaned directly. Often decisions about entering a foreign market are made after a subjective assessment of the situation.

However, there are a number of reasons why any organisation, large or small, and which is seriously intent on global trade, should consider devoting more effort to intelligence gathering. If only a few small or medium-sized enterprises made greater commitment to exporting, the impact on national finances could be significant.

The most important requirement of effective global marketing is thorough market analysis, both before entering the market and also once the market has been entered, in order to maintain continuous awareness of opportunities, threats and trends.

Specific objectives of global marketing research include the following.

- Identify attractive new markets
- Enhance profitability by pinpointing opportunities and threats
- Facilitate awareness of general market trends
- Monitor changes in customer needs and preferences
- Knowledge of competitor plans and strategies
- Identify new product opportunities in the marketplace
- Monitor political, legal, economic, social and technological trends
- Enhance the level and quality of information available for planning

It is obvious that global marketing research is more complex than domestic marketing research because of its focus on more than one country. It requires additional efforts to overcome a lack of empathy with a market and the *'foreign factor'* in general. Thus there is a need to gather information about foreign customer preferences, languages, customs, beliefs and so on which in a domestic market could be assumed to be largely known and understood.

The table below (adapted from Hollensen, 2007) summarises the principal tasks of GMR and the ways in which effective data can improve decision making.

Information for major global marketing decisions

Global marketing decision phase	Information needed
1 Deciding whether to internationalise	Assessment of global market opportunities (global demand) for the firm's products
	Commitment of the management to internationalise
	Competitiveness of the firm compared to local and international competitors
	Domestic versus international market opportunities
2 Deciding which markets to enter	Ranking of world markets according to market potential of countries/regions
	Local competition
	Political risks
	Trade barriers
	Cultural/psychic 'distance' to potential market

Global marketing decision phase	Information needed
3 Deciding how to enter foreign markets	Nature of the product (standard versus complex product)
	Size of markets/segments
	Behaviour of potential intermediaries
	Behaviour of local competition
	Transport costs
	Government requirements
4 Designing the global marketing programme	Buyer behaviour
	Competitive practice
	Available distribution channels
	Media and promotional channels
5 Implementing and controlling the global marketing programme	Negotiation styles in different cultures
	Sales by product line, sales force customer type and country/region
	Contribution margins
	Marketing expenses per market

2.2 Differing GMR objectives of established and new global marketers

To an organisation whose horizons have so far been limited to its domestic market, all global marketing research is a new activity. Since it has no experience in foreign markets it needs information to take fundamental decisions. These may be summarised as four questions.

(a) **Should the organisation look to foreign markets at all?** This will involve an assessment of overseas market demand and the organisation's potential share in it compared to domestic opportunities.

(b) **Which market(s) should it enter?** Potential foreign markets need to be ranked according to size, competition, investment and risk.

(c) **How should the organisation enter the selected markets?** Detailed analysis is required of market size, global trade barriers, transport systems and costs, local competition, government regulations and political stability.

(d) **What marketing programme is best suited to the selected market?** For each selected market a detailed knowledge of buyer behaviour, competitive practices, distribution channels, promotional methods and media will be required.

2.3 Undertaking research

In considering researching foreign markets, an organisation has a number of options.

- Do no research at all
- Use feedback and assistance from the distribution channel
- Employ an global research agency
- Conduct the research 'in-house'

There are a number of criteria that will have a bearing on the decision.

- The company's resources
- The company's experience in global trade
- The amount of foreign trade as a proportion of turnover
- The degree to which the foreign trade is ongoing

If the answer to all the above questions is '*very little*' then it is probably best that the organisation puts little of its own resources into researching the market, since the return on investment is likely to be poor.

2.3.1 The small organisation entering a foreign market for the first time

Many small organisations do not carry out any market research at all. There are a variety of reasons.

(a)   Small organisations generally use **indirect means** of exporting, and hence tend to rely on the channel intermediary for market intelligence.

(b)   They often lack the available **resources** to devote to GMR.

(c)   The **amount of foreign business** does not warrant significant expenditure on market research.

The lack of market knowledge, and reliance on the expertise of an intermediary, will limit the ability of the organisation to expand. For this reason most countries, including the UK, have **sources of low cost advice** and assistance targeted at the small business to encourage the organisation to develop and expand global trading operations.

2.3.2 The medium-sized organisation wishing to extend its export business

An organisation in this situation will be faced with problems of finding new markets, representation, and distribution. There are several options.

*   Use existing channels of distribution to obtain intelligence
*   Use available assistance and advice services
*   Engage a GMR agency

2.3.3 The larger organisation with an overseas presence wishing to increase its foreign market operations

Larger companies

*   Tend to be exporting on a regular basis
*   Gain between 10% to 30% turnover from export operations
*   Have some 'In house' expertise, usually In the form of an export department

2.3.4 Major multinationals introducing new products

The multinational generally tries to introduce products that have a global appeal. Generally such products will be either made under contract, licensed, or made by a subsidiary. Local presence means that the multinational can call upon the expert resources of its own employees in that market. All the GMR activities are usually conducted 'in house'.

Activity 1

With reference to the five phases of global marketing decision-making identify at least five key pieces of information that your organisation requires in order to make effective decisions regarding the country to enter.

3 Information strategy and sources

3.1 Why organisations need an information systems strategy

> **Definition**
>
> The **information systems (IS) strategy** refers to the long-term plan concerned with exploiting IS and IT either to support business strategies or create new strategic options.

A proper strategy for information systems is justified on the grounds that it:

- Involves high costs
- Is critical to success for many organisations
- Is now used as part of the commercial strategy in the battle for competitive advantage
- Impacts on customer service
- Affects all levels of management
- Affects the way management information is created and presented
- Requires effective management to obtain the maximum benefit
- Involves many stakeholders inside and outside the organisation

It is now recognised that information can be used as a source of competitive advantage. Many organisations have recognised the importance of information and developed an information strategy.

IT is an enabling technology, and can produce dramatic changes in individual businesses and whole industries. For example, the deregulation of the airline industry encouraged the growth of computerised seat reservation systems. IT can be both a cause of major changes in doing business and a response to them.

Parties interested in an organisation's use of IT are as follows.

(a) Other business users – for example to facilitate Electronic Data Interchange (EDI).

(b) Governments – eg telecommunications regulation, regulation of electronic commerce.

(c) IT manufacturers looking for new markets and product development. User/groups may be able to influence software producers.

(d) Consumers – for example as reassurance that product quality is high, consumers may also be interested if information is provided via the Internet.

(e) Employees – as IT affects work practices.

3.2 Information systems and corporate/business strategy

It is widely accepted that an organisation's information system should support corporate and business strategy. In some circumstances an information system may have a greater influence and actually help determine corporate/business strategy. For example:

(a) IS/IT may provide a possible source of competitive advantage. This could involve new technology not yet available to others or simply using existing technology in a different way.

(b) The information system may help in formulating business strategy by providing information from internal and external sources.

(c) Developments in IT may provide **new channels** for distributing and collecting information, and/or for conducting transactions eg the Internet.

Some common ways in which IS/IT have had a major impact on organisations are explained below.

(a) The type of products or services that are made and sold. For example, consumer markets have seen the emergence of home computers, compact discs and satellite dishes for receiving satellite TV; industrial markets have seen the emergence of custom-built microchips, robots and local area networks for office information systems. Technological changes can be relatively minor, such as the introduction of tennis and squash rackets with graphite frames, fluoride toothpaste and turbo-powered car engines.

(b) The way in which services are provided. High-street banks encourage customers to use 'hole-in-the-wall' cash dispensers, or telephone or Internet banking. Most larger shops now use computerised Point of Sale terminals at cash desks. Many organisations use e-commerce: selling products and services over the Internet.

(c) The way in which markets are identified. Database systems make it much easier to analyse the market place.

(d) The means and extent of communications with customers.

Before going on to consider the GMR process, it is important to consider the types of data source available to the global marketer.

3.3 Human sources of information

Human sources provide the largest source of information to global marketers. For well established global marketing companies, the principal human information source is the managers of subsidiaries, branches and associates abroad.

Not only do they live in the foreign cultural environments but they also appreciate the company's business objectives.

- They can distinguish relevant from irrelevant information
- They can use personal contacts to acquire unpublished information

Managerial colleagues do not represent the sole source of human information. Consumers, customers, distributors, suppliers and even competitors are all important sources of information. These groups are particularly valuable in providing marketing assessments from their own particular standpoint.

Collectively they may give a balanced view of a market, its potential and its problems. Friends, acquaintances, consultants and professional colleagues can provide 'headquarters' marketing managers with information that is reliable and objective.

3.4 Documentary sources

Documentary sources are compiled without any knowledge of the reader or the precise purposes for which the information is required.

The biggest problem in using documentary sources is that there are simply so many of them, while none may specifically address the point of interest. For example, a marketing manager interested in selling shoe laces in India is unlikely to find written material on such a market, but may have access to numerous pieces on India and on the shoe trade. Without other information sources it would be dangerous to conclude that because say, ninety per cent of the population wore shoes there must be a large market for shoe laces.

3.5 Direct sources

Direct sources are those from which the marketing manager derives information without any intermediate analysis that could reduce its levels of accuracy and relevance. There are three general types.

(a) **Direct observation and specialist knowledge**. For example, in a tour around a plastics factory the marketing manager might be shown a new lightweight durable plastic. He knows that a persistent

problem with his kitchen appliances has been that they are too heavy to be carried easily, and sees in the plastic a product that might reduce the weight of some of the motor's metal components without losing durability.

(b) **Direct observation and background information.** For example, a global marketing manager based in the UK may have heard of the French hypermarket, and received written reports about them, but only by visiting one can he properly appreciate the ambience they offer to the customer.

(c) **Personal experience supporting indirect information.** For example, a marketing manager considering the potential of the Norwegian market for the organisation's product will note from a map that Norway is a geographically dispersed country with rugged terrain. But a plane trip from Oslo to Tromso will indicate just how rugged is the terrain and drive home the point that Norway extends for over 1,500 miles from north to south. Such information should encourage particular attention to distribution plans.

3.6 Sources of information and assistance to organisations

Organisations are able to obtain considerable help and advice in their attempts to research foreign markets. Contacts may be made with a variety of organisations, including the following.

- Overseas agents and distributors
- Department for Business Innovation and Skills (BIS)
- Banks
- Embassies and consulates
- Trade and professional organisations
- Academic institutions
- Chambers of Commerce

Various global bodies including the United Nations, the European Union, the Organisation for Economic Co-operation and Development and the Global Monetary Fund produce statistical publications on a variety of areas.

(a) **Foreign governments** and global bodies (such as the United Nations) publish reports and statistics

(b) Foreign **trade associations** and industry bodies publish surveys. Many also have global links

(c) Global news and information **agencies** (such as Reuters) publish **world-wide reports**

(d) **Global newspapers** and journals are available (in print, and frequently also on the Internet)

(e) **Internet sites** give access to foreign educational and governmental institutions and their databases, and to the websites of commercial organisations around the world. Access is not restricted to working hours (which helps when there are time differences) and information can be downloaded for later translation.

3.7 Documentation

The red tape of global marketing requires that a great deal of information has to be processed by a wide range of people working in different languages with different systems. Mistakes in the information provided by exporters and a lack of management attention often cause problems.

- Relationships with overseas customers are harmed by delays
- Orders and profits may be reduced or lost

Following a year of consultation with business and analysis of the international trade environment in the UK, the UK Government published an action plan for Simplifying Trade cross UK Borders. SITPRO provides assistance for traders who need to understand the documentation and procedures required to trade internationally. SITPRO is working with organisations in the UK and internationally to develop policies that will make trade simpler and has designed the UK Series of Aligned Export Documents to simplify international trading paperwork.

http://www.sitpro.org.uk/

3.8 Research on the Internet

The amount of information in the world is doubling every seven years and there are now around 1 billion Internet users worldwide in more than 200 countries (World Bank, 2008). The Internet is destroying old power structures. As knowledge flows into every home and business on demand, it empowers consumers and businesses.

For many organisations the Internet provides an important medium for conducting GMR. New product concepts, advertising copy, promotional offers can all be tested and evaluated over the Internet. Global consumer panels have been created to test marketing programmes across borders. There are now at least eight different uses for the Internet in GMR:

1   Online survey and buyer panels
2   Online focus groups
3   Web visitor tracking (or web analytics)
4   Advertising measurement and tracking
5   Customer identification and registration systems
6   Email marketing lists
7   Embedded research through purchase processes
8   Observational research such as blogs, discussion forums, social networks etc

Today the real power of the Internet in GMR is the ability to easily access up-to-date secondary data. It is also increasingly used to conduct primary research. While a sample consisting solely of Internet users may exhibit some bias, as more of the general population accesses the Internet the tool will become increasingly accurate and powerful.

As you come across useful sources please share them in the discussion forum with your colleagues on the programme. Provide a short summary as to why they are useful.

4   Databases and expert systems

4.1 Databases

A management information system or database should provide managers with a useful flow of relevant information which is easy to use and easy to access.

> **Definition**
>
> In simple terms, a **database** is a large file or files of data, with the file structured in such a way that the data can be processed by different users in a larger number of different ways.

A database is, by implication, a computer file, and the collection of programs that are written to process data on the file in the many different ways is referred to as a database management system (DBMS). These systems provide managers with support for their decision making. Marketing data can be analysed using statistical techniques, in models on anything from media mix to new site locations.

(a)   Computer databases make it easier to collect and store more data/information.

(b)   Computer software allows the data to be extracted from the file and processed to provide whatever information management needs.

(c)   Developments in information technology allow businesses to have access to the databases of external organisations. Reuters, for example, provides an online information system about money market interest rates and foreign exchange rates to organisations involved in money market and foreign exchange dealings, and to the treasury departments of a large number of companies.

(d)   The growing adoption of technology at point of sale provides an invaluable source of data to both retailer and manufacturer.

Environmental data to be included in a database for global marketing planners could include the following.

(a)   **Competitive data**

- The threat of new entrants
- The threat from substitutes
- The power of buyers
- The power of suppliers
- The nature and intensity of competition
- The strategies or likely strategies of competitors

(b)   **Economic data.** Details of past growth and predictions of future growth in GDP and disposable income, the pattern of interest rates, predictions of the rate of inflation, unemployment levels and tax rates and developments in global trade.

(c)   **Political data.** The influence that the government is having on the industry.

(d)   **Legal data.** The likely implications of recent and future legislation.

(e)   **Social data.** Changing habits, attitudes, cultures and educational standards of the population as a whole, and customers in particular.

(f)   **Technological data.** Technological changes that have occurred or will occur, and the implications that these will have for the organisation.

(g)   **Geographic data.** Data about individual regions or countries, each of them potentially segments of the market with their own unique characteristics.

(h)   **Data about stakeholders in the business.** Employees, management and shareholders, the influence of each group, and what each group wants from the organisation.

In other words data which covers the key elements of the general and market environment should be included in a database for global marketing planners.

There are a number of problems associated with online databases.

(a)   Without planning and skill in carrying out online database searches, the expense can become prohibitive.

(b)   Because efficient use of online databases requires skill, a commitment to training is essential.

(c)   Online databases do not yet cover all important marketing research sources. Particular omissions are parts of the trade press and some important statistical data. Total reliance on databases and an attitude of if-it-isn't-on-the-screen-it-doesn't-exist could be dangerous.

(d)   Online database searching can yield lots of data but not much worthwhile information.

Please find below examples of some of the globally based online databases that global marketers will find useful in undertaking comparative studies or GMR.

Online databases

Company information	
Duns European Marketing Database: http://ds.datastarweb.com/ds/products/datastar/sheets/dbzz.htm	1.8 million companies in 16 countries
European Kompass: http://library.dialog.com/bluesheets/html/bl0590.html	370,000 companies
Datastream: http://online.thomsonreuters.com/datastream/	Financial data on companies worldwide

Trade data	
Textline: http://www.intracen.org/dbdir/file0019.htm	Reuters
Comtrade: http://comtrade.un.org/	UN Foreign trade database
Croners: http://www.croner.co.uk/croner/jsp/CronerHome.do?cache=false&channelId=-238414	Wide range of data on EC/USA/JAPAN
IMF/World Bank/UN: http://www.imf.org/external/index.htm http://www.worldbank.org/ http://www.un.org/	World trade statistics

Market information	
Business Monitor International: http://www.businessmonitor.com/	Market forecasts worldwide
Euro Monitor: http://www.euromonitor.com/	Covers market reports on 100 consumer markets, www.euromonitor.com/World_Consumer_Lifestyles_covers comparable country lifestyle_data across 71 countries
Profound http://www.profound.com/Default.aspx?AspxAutoDetectCookieSupport=1	Full text market research reports
Mintel: http://www.mintel.com/	Market reports

4.2 Expert systems

Expert systems are computer programs which allow users to benefit from expert knowledge, information and advice.

An expert system is therefore a programme for which the master/reference file holds a large amount of **specialised data**. The user keys in certain facts and the program uses its information on file to produce a decision about something on which an expert's decision would normally be required.

Expert systems can give factual answers to specific queries, but they can also indicate to the user what a decision ought to be in a particular situation, and in this respect, expert systems can be a form of decision support system for managers. Applications of expert systems include the following.

(a)   In some database systems, to speed up the process of retrieving data from a database file

(b)   In diagnostic systems, to identify causes of problems, such as in production control systems

(c)   Global tax advice

(d)   Forecasting of economic or financial developments, or of market and customer behaviour

(e)   Project management

(f)   Surveillance, for example of the number of customers entering a supermarket, to decide what shelves need restocking and when more checkouts need to be opened.

4.3 Expert systems and marketing

In theory, expert systems are essential to today's goal of **precision marketing** (the mass-customisation model in which the marketing approach is matched precisely to the needs of the individual).

Precision marketing is problematic because of the difficulty of manipulating the vast quantities of data involved. Until recently, even starting on such a task was near impossible; nowadays computers can easily handle the 'paperwork' involved, although they cannot so easily take decisions without being fed very elaborate sets of rules to govern every possible situation.

The expert system is the longer-term solution to this kind of dilemma. With an expert system, the computer can be taught how to make the necessary decisions using artificial intelligence. It may even 'learn from experience' in some circumstances.

Expert systems have been used to filter marketing data for quality problems, to develop forecasts, to set sales targets and to train sales teams. The problem to be solved must be clearly defined and narrow in scope, and the benefits of developing the Expert System should outweigh the costs.

Other marketing related expert systems have been developed to

*   Set marketing objectives
*   Select advertising strategies
*   Recommend promotions
*   Filter new product ideas

Many companies have become marketing household names on the basis of their marketing databases, namely, Experian of the UK, with its Mosaic software, and CACI of the US/UK with its Acorn software. MOSAIC and ACORN databases are designed to assist in the market segmentation process and are derived from comprehensive geographic (location) and demographic data. Both of these software shells have been adapted to suit the data available for many of the countries of the world and provide global marketers with geo-demographic databases.

5 The GMR process

A diagram of the process of global market research is shown below.

Stage 1: monitoring global markets

Monitoring is by far the most widespread activity in global marketing research. It involves passive information gathering in which the organisation has identified a particular market on which information needs to be collected but, as yet, does not warrant active measures.

If the information that has been acquired indicates the potential for a new market, the manager will proceed to the next stage, investigation.

Stage 2: investigation

Market opportunity

The most important aspect of the investigation stage is the accurate assessment of market opportunity.

(a)   Existing demand concerns current purchases of the type of product. Often, it is this demand that originally attracted the company's management to the market during the 'monitoring phase'. The level of existing demand alone may be enough to justify entry to that market, to win a share.

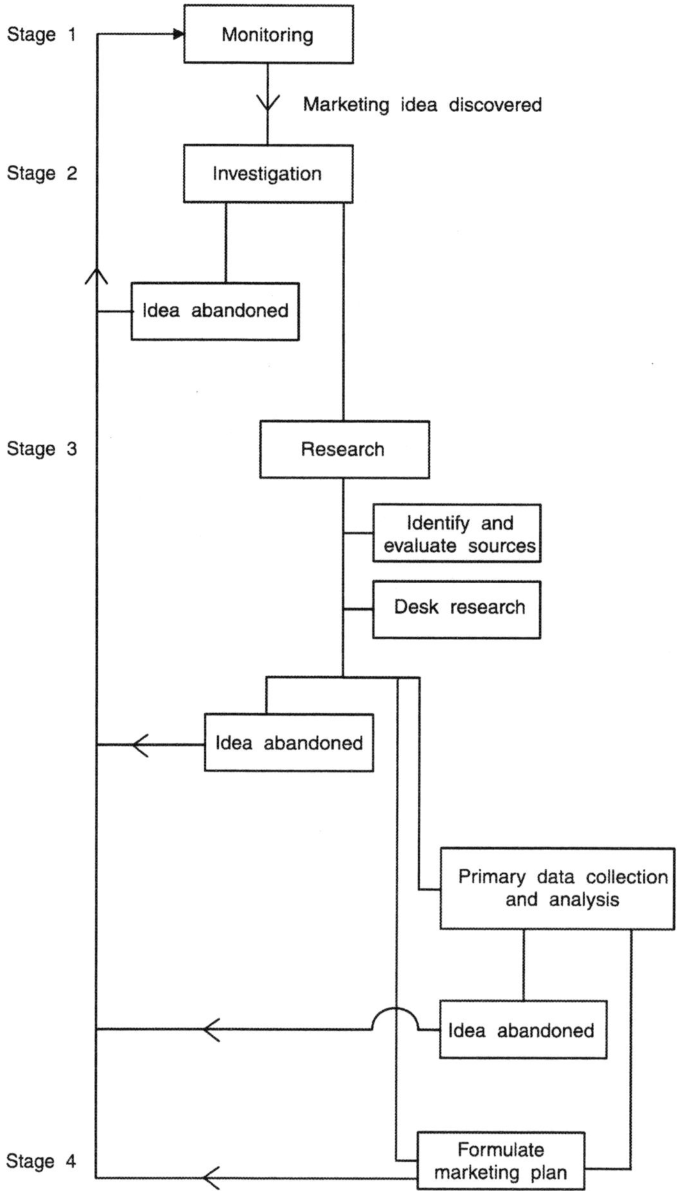

(b)   Latent demand is demand which exists in the market but is currently untapped because of a defect in the marketing mix used by existing suppliers. Often it is simply a question of price but it may be that the distribution channels are not adequate or that the product does not offer features which that market particularly values.

(c)   Incipient demand is the demand that depends on trends. It is used most often by manufacturers of goods that require wealth levels in an economy to be above a certain minimum before the volume of demand becomes sufficiently great for the market to become profitable.

Stage 3: research

By this stage of the GMR process the new marketing idea has to look sufficiently promising to be worth committing a significant budget to. Whereas during the investigation stage enquiries are often informal and information may be collected as part of other activities, market research proper is proactive. In the global context, this can be expensive.

5.1 Planning

A vital part in any research exercise is to plan it properly. Although no entirely standard plan can be drawn up, it is likely to have the following elements.

(a)   **Define the scope of the project**

Identify the geographical and political market(s) of interest, the market segments and, not least, the end objective.

(b)   **Define the project's information needs**

There must always be a balance struck between the use of information and the cost and difficulty of obtaining it. Posing the following questions will help in finding the correct balance.

(i)   **Why** is the information needed?

(ii)   If it is obtained, **how** will it be used?

(iii)   **Where** may information be obtained from? Is it available directly as a secondary data source, capable of production by analysis of existing secondary data, or does it require primary data collection?

(iv)   **What is the information worth** in financial terms?

(v)   What is the **cost of not obtaining it**?

(c)   **Evaluate the available sources for the required information**

It is no use for the global marketer to assume that because data is available it will necessarily be valuable! Often great caution is needed.

(d)   **Undertake the desk research and evaluate its findings**

Desk research, using documentary sources, should always be done before conducting or commissioning any primary data collection. It is almost invariably cheaper and, if its results are disappointing, the project may be abandoned without any primary data collection being undertaken at all. Even if the project does continue, and primary data is needed, careful analysis of the secondary data can focus attention on the exact primary data requirement and so reduce the costs of the research exercise.

(e)   **Undertake field research**

The collection of primary data is sometimes known as field research. Field research is carried out 'on the spot' in a number of areas, notably customer research, advertising research, product research, packaging research and distribution research. Techniques involved in the collection and analysis of primary data are as follows.

• Experimentation	• Consumer panels
• Sampling	• Trade audits, such as retail audits
• Piloting	• Pre-tests
• Observation	• Post-tests
• Questionnaires	• Attitude scales

(f)   **Knowledge of the culture** of a society is clearly of value to business. Marketers can adapt their products and appeal according to the culture of their intended export market. Multinational companies often establish new subsidiaries in a different country. Much was made of the cultural difficulties which Nissan was supposed to encounter when investing the US and UK. Nissan had to teach its new recruits about the culture of the company.

Whether you love it or despise it, it is no fluke that McDonald's has successfully infiltrated the Indian market. According to branding expert Preeti Chaturvedi. Chaturvedi points out that while countless fast food franchises have attempted to open in the sub-continent, McDonald's has not only lasted the distance, it's branching out and is opening more outlets due to its effective marketing strategies. "*When McDonald's India launched in 1996, urban Indians in Mumbai and Delhi typically ate out three to five times a month, according to AT Kearney, the management consultancy. In the 12 years since then, that average frequency has doubled and analysts forecast that by 2011 the Indian quick service restaurant market will be worth about $6.3 billion.*" The first Indian McDonald's outlet opened in Mumbai in 1996. Since then, outlets have begun trading in metropolitan and bigger towns across the country. Amit Jatia, managing director of McDonald's India, says "*The past decade has witnessed a marked change in Indian consumption patterns, especially in terms of food. Households in middle, upper, and high-income categories now have higher disposable income per member and a propensity to spend more.*" The interest of fast food multi-nationals in the Eastern regions is well documented. Warren Lui's recent book, *KFC in China*, published by John Wiley & Sons, follows the rise of the chicken giant in a country that has heavily restricted international investment.

Preeti Chaturvedi:http://ipm.ge/article/How%20McDonald's%20evolved%20its%20marketing%20in%20India_ENG.pdf

5.2 Desk research requirements

Doole and Lowe (2008) suggest a '12C' checklist for information which a marketing information system for global markets should contain. This is shown below.

The 12C framework for analysing international markets		
Country	**Consumption**	**Communication**
General country information	Demand and end use analysis of economic sectors that use the product	Promotion
Basic SLEPT data		Media infrastructure and availability
Impact of environmental dimensions	Market share by demand sector	Which marketing approaches are effective
	Growth patterns of sectors	
	Evaluation of the threat of substitute products	Cost of promotion
		Common selling practices
		Media information
Concentration	**Contractual obligations**	**Capacity to pay**
Structure of the market segments	Business practices	Pricing
Geographical spread	Insurance	Extrapolation of pricing to examine trends
	Legal obligations	Culture of pricing
		Conditions of payment
		Insurance terms

Culture/consumer behaviour	**Commitment**	**Currency**
Characteristics of the country	Access to market	Stability
Diversity of cultural groupings	Trade incentives and barriers	Restrictions
Nature of decision-making	Custom tariffs	Exchange controls
Major influences of purchasing behaviour		

Choices	**Channels**	**Caveats**
Analysis of supply	Purchasing behaviour	Factors to beware of
International and external competition	Capabilities of intermediaries	
	Coverage of distribution costs	
Characteristics of competitors	Physical distribution	
Import analysis	Infrastructure	
Competitive strengths and weaknesses	Size and grade of products purchased	

Desk research involves work in two main areas.

(a) Collecting information

- The business environment as a whole (general background analysis)
- The country's economic structure of particular relevance to imports (market access analysis)

(b) Beyond this general level, desk research is concerned with obtaining as much information as possible on the structure of the market for the company's goods, the practices of the market and an analysis of competitor's positions. Each of these is considered below.

5.3 General background analysis

It is important that the organisation has a sound general knowledge of the country. This involves analysing various data.

(a) **Geography**: location, size, topography, climate and so on.

(b) **Population**: size, distribution by location, race, religion, income, education, age, gender etc.

(c) **Language**: official, business and other indigenous languages.

(d) **Government**: type of constitution, roles of central and local government, political climate and attitudes to foreign trade, economic and social policies.

(e) **Basic economic data**

(i) Indicators: currency exchange rates, balance of payments, foreign currency reserves, debt situation, GNP, GDP, national income, per capita income and price inflation indices.

(ii) Structure of the economy: rate and distribution of employment, output of the various industrial, service and extractive sectors.

(iii) Economic Development Plans: funds allocated, time horizon, target sectoral allocations etc.

(iv) Policy relating to foreign investment: both inward and outward.

(v) Budgetary provisions.

(f) **Infrastructure**: seaports, airports, roads, rail, postal system, telecommunication system etc.

(g)   **Foreign trade data**

   (i)   Total foreign trade: exports, imports, balance of trade, and relative importance to economy.
   (ii)   Main export products: volume, value and destination.
   (iii)   Main import products: volume, value and source.
   (iv)   Trade between target country and the UK: exports and imports.
   (v)   Multilateral trade agreements affecting trade between the target country and the UK.

(h)   **Trading systems**

   (i)   Nationalised: controlling authority, central buying agency, payment procedures etc.
   (ii)   Private sector: Chambers of Commerce, Trade Associations, agents, distributors etc.

5.4 Market access analysis

GMR is also concerned with the governmental policies and regulations that specifically affect market entry. These are important because in most countries there are significant variations in the attitude towards, and the regulations imposed upon, a particular importer and the way in which they may be permitted to import. The research may be under various headings.

(a)   **General import policy**

   (i)   Membership of customs union, free trade area, WTO etc.

   (ii)   Special trade relationships between the trading countries involved (for example between the UK and Commonwealth countries).

(b)   **Restrictive import licensing regulations**: categories, conditions for acquisition, procedures etc.

(c)   **Import tariff system**: classification system, tariff rates, bases of assessment for duty etc.

(d)   **Other factors**: foreign exchange controls, import deposit schemes, anti dumping and minimum price regulations; food, health and safety regulations; selling, promotion, packaging and labelling restrictions; patents and trademark protection in that country; fair trade and anti trust rules; taxation and shipping documents required.

(e)   **Ownership of trading companies** by foreign nationals: restrictions, regulations and imposition of local ownership.

5.5 Market structure

In order to trade in a foreign market it is vital to have a detailed knowledge of the market structure and behaviour. Knowledge of the composition, profile and behaviour of the market is fundamental to marketing, whether domestic or globally based.

Typically in an global context, information is required on the following points.

(a)   Who are the **main customers**? How, why, where, when and how much do they buy?

(b)   What are the **main channels of distribution** used in such markets? Who are the main distributors, agents etc? Do reciprocal or other possibly limiting trading practices exist that could hamper market entry?

(c)   What **product attributes**, specifications and developments are there in the market?

(d)   Who are the **main suppliers** to the market? What are their relative positions, shares, strengths and weaknesses, strategies and performance?

(e)   Are there significant **geographic variations** in customer requirements, distribution costs, product use, and promotional needs?

(f)   What **facilities** are there for promoting the product into the market? What is the effectiveness of the various forms of promotion and media available?

(g)   What is the **size** of the total market and potential target segments? How durable is the market? For example is it likely to disappear in adverse economic or political conditions?

5.6 Competitor analysis

World market competition is intensifying in most products and markets. If a company is interested in anything more than an insignificant niche in world trade it must have an extensive knowledge of competitors' plans. An organisation should be aware of the following.

(a)   Major competitors: their number, size, market shares and nationalities.

(b)   Do the major competitors have full market coverage? Do they instead specialise in certain geographic areas or market segments? Do their product ranges contain gaps?

5.7 Market practices analysis

It is very rare that the marketing mix relevant to one country can be imported unadapted to another. Hence the global marketer needs to assess the relevant aspects of the mix suitable for the target country. These include the following.

(a)   **Transport facilities**: types, prices, reliability and risks (both within and between the relevant countries).

(b)   **Distribution channels**: relative costs and benefits; possible alternatives; channel norms and trading behaviour.

(c)   **Pricing strategy**: upper and lower limits, competitor prices compared to product features and appeal, discount structures etc.

(d)   **Promotional factors**: methods normally used in the market, levels of expenditure and availability.

(e)   **Product and services**: required features and facilities before, during and after sale. Warranties, information, credit etc as forms of product enhancement.

Activity 3

With reference to the GMR planning process, prepare an outline marketing research plan for your organisation to enable it to understand more about the country you are recommending it enter.

6 The uses of GMR data

Many organisations engaged in GMR seek to spread their activities over a number of countries, and so one purpose of GMR must be to ensure that strategic positioning of products and the appropriate marketing mixes are duly identified.

GMR data can therefore be used in the following ways to identify market opportunities.

(a)   To estimate product patterns of demand/or consumption in individual markets.
(b)   To compare patterns of demand or consumption in different markets.
(c)   To identify clusters of markets with similar characteristics, which can be targeted with similar mix.
(d)   To identify strategically equivalent segments, across country boundaries.

6.1 Predicting patterns of demand

Patterns of demand change over time, for many reasons, such as changes in customer needs and expectations on the one hand, or radical product innovations on the other. GMR can identify trends in demand both directly, and also by inference.

6.1.1 Demand pattern analysis

Demand pattern analysis involves analysing production patterns (as a surrogate for demand) over time. It helps in identifying general marketing opportunities and, because it uses production statistics, it can indicate potential markets.

6.1.2 Income elasticity

We can also estimate effective demand from how much people earn. Income elasticity studies are concerned with specific products. Income elasticity describes the relative changes in demand for a product with changes in income levels. Income elasticity is expressed as:

$$\frac{\text{Change in product demand} \div \text{Product demand}}{\text{Change in income} \div \text{Income}}$$

(a)   Some goods are income inelastic. As people get richer, demand for such goods does not increase proportionately. For example, poor people might spend over 50% of their income on basic foodstuffs. In wealthy countries food accounts for about 11%.

(b)   Goods which are income elastic are very sensitive to changes in people's income. These include leisure products, some consumer durables and so on.

6.1.3 Multiple factor indices

It is commonly the case that data of the kind required for direct analysis of a market is unavailable and thus the market researcher must find alternatives. Multiple factor analysis attempts to surmount this problem by using surrogate variables that can be closely correlated with demand for the company's own product.

There is clearly much scope for error, but this may be reduced by using variables that are as close as possible to demand for the product. So whereas, for example, growth in total population might be found to have a good statistical correlation with demand for iPods, changes in population in the 15-30 age group is likely to be a better indicator. Changes in personal disposable income may be a better indicator again.

6.1.4 Regression analysis

Regression analysis takes the ideas of multiple factor indices further and adds some statistical rigour. The usual form the technique takes is linear regression, where change in an independent variable (often GNP per head) seeks to explain change in the dependent variable (normally demand for the product in question).

Care must be taken to use this technique within its limitations. In particular, the predictive ability of the regression may be limited to a narrow band of change in the independent variable. Taking the example of radio set ownership, one might find the situation shown overleaf.

The regression line can be extended at will but the growth rate of ownership of radios may fall above a certain income level. This is to be expected by intuition. As average wealth increases there comes a point when the majority of people who wish to own a radio have become rich enough to buy one and thus further increases in average wealth have less and less impact on the numbers owning them.

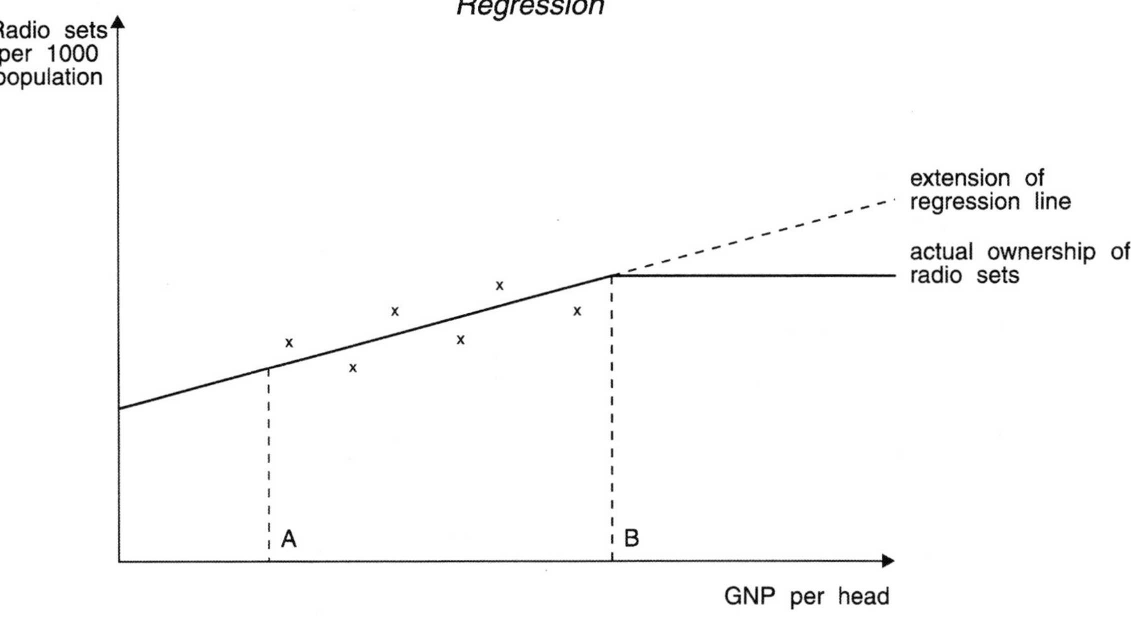

6.2 Comparing patterns of demand/consumption in different markets

6.2.1 Comparative analysis

Comparative analysis assumes that if market potential is equivalent in two markets then marketing performance should be comparable too. It usually takes the form of comparison of the same company's performance in two (or more) national markets. If the economies are broadly similar but performance is not (or vice versa) it suggests that marketing performance is not being optimised and thus raises questions to be addressed.

Factors in a comparative analysis include many of the SLEPT factors.

- Size of population
- Age structure
- Geographic distribution
- Incomes
- Lifestyles
- Communications
- Ability to pay
- Barriers to entry

6.2.2 Inter market timing differences

This technique uses the premise that certain markets have similar demand patterns for similar goods but that one leads and the other lags. Clearly this technique is useful only for comparing countries that can be assumed to have similar economic, social and cultural conditions. The diagram below shows the similarity in the rate of acquisition of television in the 1950s and 1960s between the UK and West Germany.

At its simplest, the technique enabled the prediction of growth or decline in the then West German market by taking that of the UK market a few years earlier and adjusting for the different numbers of households in the two markets.

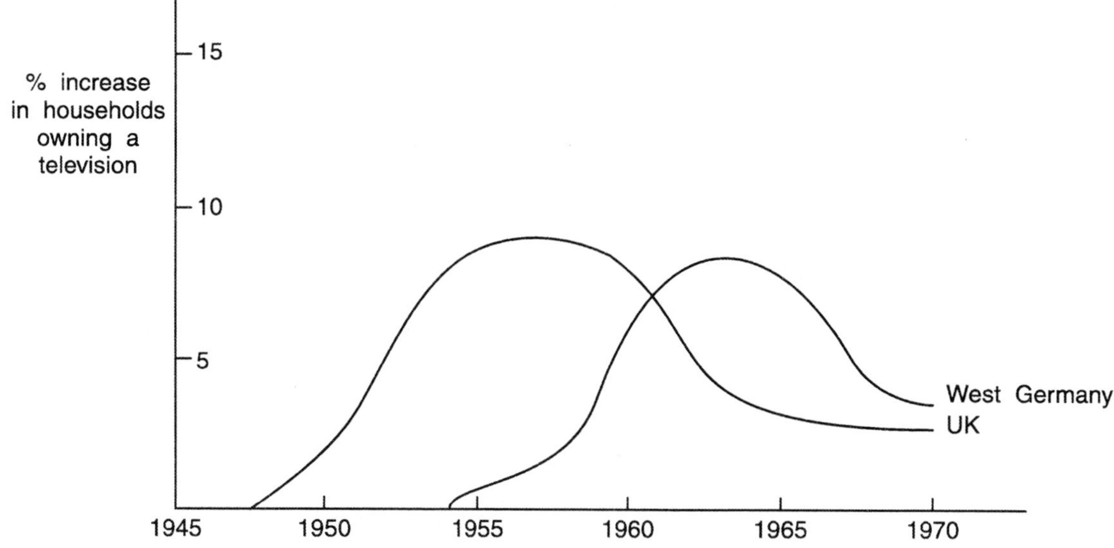

6.2.3 Identifying clusters of markets with similar characteristics

A logical extension of using one market to predict the behaviour of another is to identify clusters of markets with the same characteristics. Put another way, this is rather like segmenting groups of customers – instead the countries of the world are being segmented. Cluster analysis involves mathematical techniques to identify similar markets.

(a)   According to Sethi (1971) four sets of variables can be used to compare between countries:

- Production and transportation
- Consumption (income, eg number of cars)
- Trade data derived from import or export figures
- Health and education

Each of these were scored within Sethi's (1971) research and countries were fitted into one of seven groups, with the implication that the similarity of countries within a group was strong enough to justify similar marketing approaches.

(b)   Business Global identified indices covering market size (eg population), market growth rates and market intensity (relative concentration of wealth and purchasing power). For example the former Soviet Union, Poland and Brazil occupied one cluster, having enjoyed similar growth rates between 1986 and 1990 and having a similar 'intensity'.

6.2.4 Identifying strategically equivalent segments

In addition to identifying clusters of countries, it is possible that an organisation may wish to segment each market and pursue these segments only.

> **Definition**
>
> **Segmentation** is the subdividing of the market into increasingly homogenous subgroups of customers, where any subgroup can be selected as a target market to be met with a distinct marketing mix.

In global marketing, as in domestic marketing, organisations can use more than one segmentation variable, a primary segmentation variable and a secondary segmentation variable.

For example, for wet shaving equipment, the primary segmentation variable might be gender. For example Gillette produces different razor blade holders for men and women. These markets will indicate the best opportunities to establish a permanent presence. Secondary segmentation variables might be age or

lifestyle. Secondary markets might be handled in a more pragmatic way and comprehensive marketing research would be required to understand more about the opportunity.

In global marketing, using a country as the primary segmentation variable would seem common sense, certainly for consumer products, given the cultural influences, political risk factors and so forth identified earlier. If an organisation pursues a multi-domestic strategy it will almost certainly use country as the primary segmentation variable. Even a global organisation to whom the world is a single market, will develop country-based segments in consumer markets.

This leads us to the notion of **strategically equivalent segments**, which transcend national boundaries. Here are some possible examples.

(a) In South East Asia, people of Chinese ethnic extraction form the business elite in many countries (eg Malaysia) despite government attempts to reduce imbalances. Therefore, for certain types of goods and services, overseas Chinese (eg in Malaysia, Singapore, Thailand etc) could be considered a strategically equivalent segment.

(b) Geo-demographic segmentation can also be used. Experian analyses ten classifications of European consumers as shown below.

• Wealthy suburbs	• Industrial communities
• Average-income areas	• Dynamic families
• Flats – luxury	• Low income families
• Low income inner city	• Farming/rural
• Municipal/social housing estates	• Holiday/retirement areas

Such worries may not exist for many industrial markets, where there are fewer buyers, and where purchase decisions are in theory taken according to rational criteria. The market for aircraft engines comprises the world's airlines, and it may not be possible or necessary to segment this market.

7 Some problems in GMR

7.1 Researching numerous markets

In the initial stages, global marketing research may address a hundred or more markets in the broad scan for market opportunities, and certainly concentrate on a minimum of a handful of markets as enquiries become more focused. This generates a number of problems.

(a) **Cost versus profit** considerations. Researching multiple markets can give rise to economies of scale and experience but even so a marginal cost is incurred for each. Moreover many markets are small, heterogeneous and with limited profit prospects. This then justifies only moderate market research expenditure and has forced GMR practitioners to develop methods and techniques geared to limited budgets. It also requires the researcher to exercise judgement as to which countries would be worth investigating. Hence the importance of informal methods of research at the initial stages.

(b) **Differences between countries** gives rise to several problems.

   (i) **Definition**: defining the problem with respect to various countries. Bicycles are treated as recreational items in the UK, largely sporting in nature in parts of continental Europe, and as a means of transport in much of the Third World. Thus the definition of research to be carried out and the categories provided in surveys should be sufficiently broad to encompass the variety of uses, reasons and responses that might be evoked.

   (ii) **Classification**: meaning and classification problems in surveys and their analysis due to different behavioural and living patterns.

   (iii) **Frame**: sources of sampling frames from which to draw samples may not be reliable, especially in under developed countries.

(iv)   **Non response**: the desire to co-operate with a survey will largely depend on the norms of that culture. It should not be assumed that questions freely answered in one country will be accepted in another.

(v)   **Comparability** is weakened by many factors, particularly differing cultural values, consumption patterns, political influences and reliability of data sources.

7.2 Secondary data problems

There are normally three critical deficiencies regarding secondary data that are important to the global market researcher.

(a)   **Lack of data.** Without doubt, the USA has the best documented commercial information. Key economic indicators, compilations of trade statistics and the work of trade associations, state and local government, and management groups, are readily available to the market researcher. But few countries come close to matching the sheer volume and diversity of data collected in the USA. Until the United Nations began assimilating world economic data, there was often little available other than very rough estimates for many lesser developed countries.

(b)   **Comparability** and **timeliness**.

(i)   This problem is at its most acute in less economically developed countries which, despite recent attempts to remedy the situation, often choose to devote a rather lower proportion of their resources to data collection than advanced economies. As a consequence data is often many years out of date and collected on an infrequent and unpredictable schedule.

(ii)   More specific problems involve different definitions used in data collection, the fact that different base years have been used for comparative purposes and that gaps may exist. As an example a television is classified as household furniture in the USA but not in Germany (where it is treated as a recreational purchase) leading to problems of comparability of data collected in different countries.

(c)   **Lack of reliability** of some of the secondary data that is collected. Much of it has to be analysed with great care.

(i)   In many economically underdeveloped countries, national pride takes precedence over statistical accuracy.

(ii)   Moreover, companies have been known to reduce their production statistics so that they reconcile to sales reported to the tax authorities!

As a practical matter, before basing a marketing strategy on the analysis of secondary data, it is worthwhile asking the following questions.

(a)   Who collected the data? Would there be any reason for deliberately misrepresenting the facts?

(b)   For what purpose was data collected?

(c)   How was the data collected (ie what was the survey methodology used)?

(d)   Is the data consistent with one another and logical (in the light of known data sources or market factors)?

Global case study

Segmenting the Mobile Phone Market in India

Faced with diminishing marginal returns on voice revenues, Airtel was forced to abandon its one-size-fits-all demographic segmentation. It came up with a strategy where it reclassified users into different segments based on attitudes, VAS adoption traits and average revenue per person (ARPU). Airtel focused its attention on two segments in particular. The first group was known as "funsters" – consumers aged

between 18-35 years old, who share a common trait – a high adoption of VAS. Airtel believes that with some targeted marketing, spending from this group can increase quite significantly. The telco has exclusive tie-ups with application providers like Google and has identified music-on-demand as a key value proposition. The other segment of focus is the "achievers", who are the top five percent of mobile users. While this segment isn't a high adopter of VAS, it contributes to revenues nearly 10 times that of Airtel's ARPU, currently at US$10. To enhance customer experience for this segment, Airtel has separate priority relationship managers. It also partnered with HTC and Blackberry to offer high-end handsets catering specifically to this segment. Brands entering India must understand that ubiquity in diversity is an inherent dynamic of the market. Marketers must always be ready to unlearn and re-imagine their mobile strategies to succeed in India!

http://www.imediaconnection.com/content/18698.asp

18 March 2008

7.3 Problems in collecting primary data

Where the informal and secondary sources are inadequate, as a last resort the market researcher must turn to collecting primary data. This is both expensive and difficult and the task is often handed over to specialist agencies to carry out. The problems involved in primary data collection include the methods of data collection, non response, survey control, illiteracy and language.

7.3.1 Data collection

Except in the most unusual circumstances it is too time consuming and too expensive for the market researcher to survey the whole of the target population for the product under consideration (even if it can be defined adequately in the first place). Consequently a sample needs to be taken.

7.3.2 Questionnaires

Survey research, to obtain primary data, normally requires a questionnaire. There are three main characteristics of a good questionnaire.

- The questions are easy for interviewees to answer and for the interviewer to record
- It acquires the necessary information but keeps the interview to the point
- It is straightforward

In order to make sure that the questionnaire achieves these characteristics, a number of design principles need to be adhered to. The main ones are as follows.

- Questions should address one point only
- Leading questions that suggest their own answer should be avoided
- Ensure that questions mean one thing only
- Make the language in which the questions are framed easy to understand
- Closed questions make analysis faster as they list the possible replies to questions
- Avoid asking questions of a personal and potentially embarrassing kind

Assessment advice

Design a short questionnaire to help you with your assignment. Try it out on your Study buddy and ask them to give you feedback on your questionnaire design.

7.3.3 Problems in response

People's unwillingness to provide information is a feature of many countries. Several reasons can be suggested for this.

(a) Tax evasion and avoidance of responsibility. Anyone asking questions may be suspected of being a government employee.

(b) Wish to preserve secrecy. Many business managers regard data relating to their company or markets as potential help to competitors and accordingly response rates are low when they concern anything that might be regarded as an aid to a competitor.

(c) Protection of personal privacy may be a key obstacle to response.

(d) Cultural taboos and norms

    (i) Topics that are freely discussed in some countries (for example birth control methods) are taboo for discussion in others except in the most intimate situations.

    (ii) Similarly the authority to respond may not be the same in all cultures. Thus while in one country a female head of household may feel authorised to discuss a topic with a researcher, in other countries there may be a marked reluctance to give information.

In carrying out surveys, there may be limitations:

(a) There may be no suitable lists (sampling frames) from which a representative sample can be selected.

(b) In many countries there is a dearth of economic, demographic and social statistics on which to base sampling methods.

(c) Inadequate communication infrastructure can be a severe practical handicap.

(d) Low levels of literacy.

(e) Problems of language and comprehension are widespread in GMR. Differences in idiom and problems of precise translation lead to great misunderstandings. In large countries such as India the researcher has to be prepared to deal with many languages and dialects when planning a survey.

This discussion should make it clear that there is a greater necessity for care, understanding and skill in GMR data analysis.

(a) Thus the analyst requires an extensive understanding of the cultural environment of the countries being researched and a talent for interpolating data that contains gaps or has other deficiencies.

(b) A final requirement would be a healthy scepticism towards both primary and secondary data since they are very likely to be imperfect, resulting in a desire to cross check and substantiate data wherever possible.

Activity 4

Many organisations are now looking to emerging and developing markets as important aspects of their global marketing strategy. Consider the main problems associated with researching those markets and ways in which these problems can be overcome.

8 Managing and implementing the GMR process

Global marketing intelligence is crucial for a organisation's strategic decisions and for the on-going implementation and review of marketing plans. The collection of strategic intelligence has to be co-ordinated at the appropriate level in the organisation and at the appropriate time.

Not only does data have to be collected, it has to be communicated to the right area of the organisation. For decisions made by local subsidiaries, the information will be gathered and stored locally. The global headquarters, however, can operate as a clearing house for research studies, actively passing on information to those who might benefit from it.

Global head office may conduct research on its own, perhaps on issues relevant to all markets or in areas where there would be duplication if the subsidiaries did all the work themselves.

Some organisations have regional HQs as well, covering a number of countries. The advantages of one marketing research department are economies of scale, and the ability to concentrate efforts.

Co-ordinating the timeliness of information is important. After all, if a product launch is critically dependent on marketing intelligence, any delay can be costly.

An important management decision is the extent to which organisations use agencies to conduct their GMR.

8.1 Using external agencies

GMR managers face considerable problems in obtaining useful information at a reasonable cost. The manager may consider hiring an external agency to carry out GMR. Before hiring such an agency however the organisation must make three crucial decisions.

8.1.1 Whether to hire an agency or not

There are a number of factors favouring the hire of an agency.

(a)   The organisation may have little or no marketing research resources.

(b)   The organisation does have marketing research resources but these are subject to peak demands. Thus it may be more economical to hire agencies to meet peak loads than to increase the organisation's permanent resources.

(c)   The organisation may have little or no experience of the foreign market to be analysed. Here an agency with that specific market experience may be of great value.

(d)   The research is of a highly specialised kind involving skills not present in the organisation (for example behavioural studies).

(e)   The research is a 'one off' exercise and thus it would be uneconomic to develop 'in house' expertise.

(f)   The organisation anticipates language or cultural problems in carrying out the research that their own expertise cannot cope with.

(g)   Management requires an objective view, which is often difficult when its own employees are aware of the pressures within the organisation for a particular answer.

On the other hand, there are also several factors that would favour 'in house' research.

(a)   Lack of suitable marketing research agencies in the foreign markets to be investigated.

(b)   The organisation is well experienced in GMR and has a good understanding of the cultural and other conditions that might affect the research.

(c)   The relevant resources and skills already exist in the organisation

(d)   The organisation needs to develop a thorough knowledge of the market. Agencies do provide information, but they may leave the organisation without the extensive knowledge which is required to get a 'feel' for the market.

(e)   The organisation would find it difficult to brief the agency adequately on its research objectives and the technical parameters and applications of the product.

(f)   The research project is a small one, requiring few interviews and relatively little time.

(g)   It may be cheaper for the organisation to execute its own marketing research. It is quite likely that economies of scale and experience can be achieved when studies are duplicated in multiple countries.

8.1.2 Which type of agency to select

In general terms the organisation may engage any one of three types of agency to execute its GMR.

(a)   **Foreign agency domiciled in the target country**. This offers the major benefit of having a comprehensive understanding of the local market, language, culture and business environment. However, it requires costly and time consuming visits by the organisation's personnel to select and brief the agency, and then to supervise the project.

(b)   **Domestic agency sending competent staff to the target country**. This is beneficial insofar as it permits easy and cheap selection, briefing and supervision of an agency. It should also offer a high standard of marketing research skills and may be less costly than hiring a foreign agency. A serious disadvantage would arise, however, if the agency or its personnel were not in fact thoroughly familiar with the foreign market, its culture or its environment.

(c)   **Domestic agency with foreign agency subsidiaries, associates or subcontracting arrangement**. This is the synthesis of the above two alternatives. Here the domestic agency works in conjunction with its overseas associates.

   (i)   The benefit for the organisation is that it only has to select, brief and supervise the domestic agency which in turn acts on behalf of the organisation with the foreign agency. Further it is assumed that the foreign agency will have a thorough understanding of the target market, its culture and its environment.

   (ii)   The major disadvantage of such an arrangement is that the organisation has little or no control over the selection and supervision of the foreign agency. Additionally there may be some communication problems between the two agencies, and the cost of utilising two agencies may be greater than with direct hiring.

(d)   **Global agency** with a worldwide presence.

8.1.3 Selecting a particular agency

Whatever type of agency the organisation finally decides to hire, it must select an appropriate agency to carry out the research. This involves obtaining information about the various agencies to evaluate their suitability and acquiring a research proposal from each for assessment. Agencies might be invited to tender.

Briefing

Each agency approached will require certain guidelines from the organisation to ensure that the proposal which it prepares is appropriately formulated.

- A clear written statement of the research problem
- An indication of the way in which the research findings will be used
- An indication of the approximate budget
- An opportunity for the agency to discuss the proposed programme

Shortlisting

In making a shortlist, the organisation should take the following factors into account.

(a)   Evidence of suitable background and experience of the agency staff in GMR and the target markets.

(b)   Details of any specialist staff (statisticians, analysts, psychologists and so on) used by the agency.

(c)   Agency experience in relevant GMR areas, foreign markets, products and research techniques.

(d)   Quality and reliability of field operations including the selection and training of staff, levels of supervision, and control and monitoring procedures.

(e)   Where large amounts of data are to be collected, the agency's facilities for data processing, analysis and report preparation.

(f)   The financial status and reputation of the agency.

A recommendation from other satisfied users of an agency can be a significant aid in shortlisting a suitable agency. The informal network outlined earlier in this chapter can be utilised here.

Final choice

The agency selected from among the shortlisted and submitted proposals will be the one which puts forward the most appealing blend of the following:

(a)   A demonstration of a sound grasp of the research problem and its objectives.

(b)   A detailed description of the research including a statement of the scope and nature of preliminary desk research, pilot studies and qualitative research. Where quantitative research is involved, a statement of the data collection method, the population to be sampled, the sample size and the sampling technique.

(c)   A statement of the staff involved and their duties.

(d)   A statement of the total cost with a detailed breakdown of the component costs, and a reasonably detailed timetable for the research programme, including a final reporting date.

8.2 Establishing the Global Marketing Information System

Once the research has been conducted it is important to incorporate this information into effective management decision making.

> **Definition**
>
> The **Global Marketing Information System** (GMIS) is an interacting organisation of people, systems and processes designed to create a regular, continuous flow in information essential to the global marketer's problem solving and decision-making activities around the world.

The GMIS can be conceptualised as a four-stage process consisting of locating, gathering, processing and utilising information. The diagram below (adapted from Hollensen 2007) illustrates the core issues to be addressed in each of the four GMIS stages.

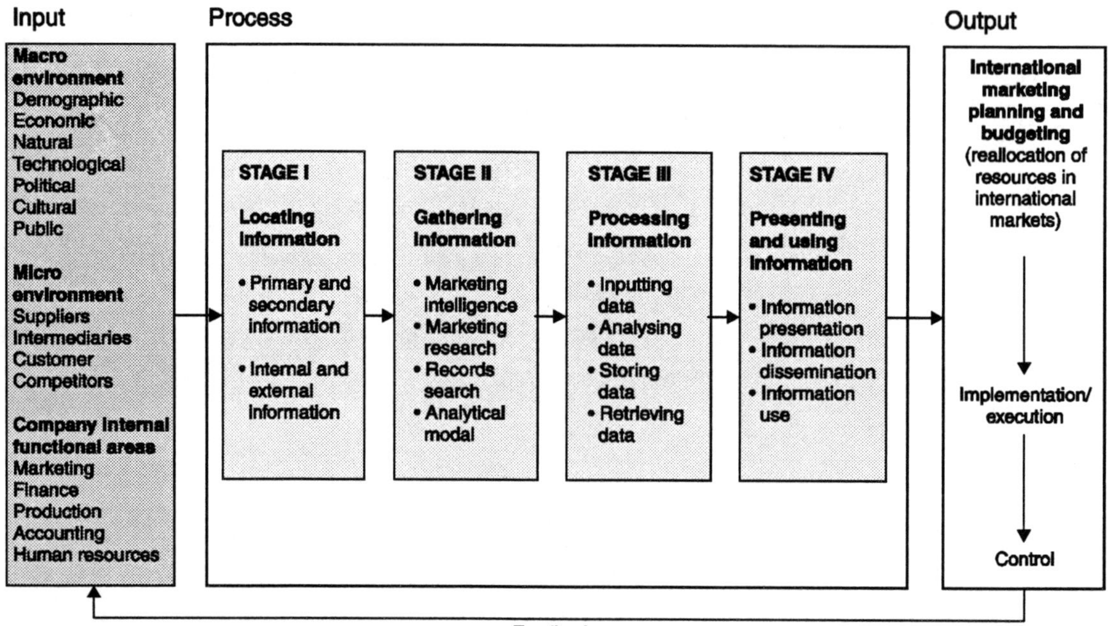

Source: Schmidt and Hollensen **(2006)**, p. 587

In this model, data is inputted from three major sources and the output is made available for analysis, planning, implementation and control purposes. Against the backdrop of a increasingly dynamic global business environment organisations are increasingly developing their GMIS systems to provide managers with real-time market information, as customers worldwide become ever more knowledgeable and sophisticated. It is increasingly important that senior decision makers and global marketers be closely involved in the fieldwork of seeing and hearing the voice of the global consumer.

Activity 5

Explore the reasons for your organisation using a marketing information system and the main types of information you would expect it to use.

Developing Global Marketing Strategy

1 Understand the role of global marketing research

- Global marketing research is more complex than domestic marketing research, as there are additional factors to consider. Some foreign markets may be attractive commercially, but there may be legal barriers to entry.

- GMR information can be used to predict patterns of demand, to compare different markets, to cluster countries and to identify strategically equivalent segments.

2 Evaluate information strategy and sources

- GMR is made difficult by the existence of different cultural assumptions in the foreign market, which may distort the results of research designed for use in the home country. Moreover, secondary data may be non-existent and unreliable.

3 Understand databases and expert systems

- Databases and expert systems can provide useful information. Expert systems in particular represent a leap in information effectiveness because of their ability to make 'decisions' and promote precision marketing.

- Cross-cultural differences must be recognised when undertaking GMR. The commissioning of a research agency experienced in the overseas market might be advisable, although the market knowledge provided may not be less thorough than it would be if the firm did the research itself.

- The organisation should establish a global marketing information system to support their global managers in making effective decisions. The consequence of poor information is significant in terms of both missed opportunities and the failure to meet existing and developing market demand.

4 Appreciate the GMR process

- There are three broad stages associated with GMR. First, is monitoring global markets. Second, comes investigation. Third is planning. All stages should feed into the marketing plan.

5 Understand the uses of GMR data

- GMR is used to identify market opportunities.

- Predicting demand patterns utilises a range of statistical tools.

- Comparing patterns of demand between markets leads to the identification of clusters of markets with similar characteristics and strategic equivalents.

6 Appreciate some problems in GMR

- However every global marketer should also bear in mind that no amount of data can be a substitute for sound judgment.

- Problems with secondary data include lack of data, reliability, comparability and timeliness.

- Problems with primary data include those associated with the data collection tools and problems in respondent response (or lack of!).

7 Understand how to manage and implement the GMR process

- Some multinational organisations centralise global research.

- Decisions about whether to hire a market research agency need to be made, again these may be local or global based organisations.

1 This answer will largely depend upon your organisation. Utilise Hollensen's model for structuring your thinking and for identifying the gaps in knowledge within the organisation.

2 This answer will depend upon your research.

3 While there is no prescribed planning process the one proposed in this chapter is a good template to follow. Define the scope of the project, define the project's information needs, evaluate the available sources for the required information and then undertake the desk research and evaluate its findings. Based upon this process you will then establish whether you need to undertake further primary research.

4 There are problems associated with researching emerging markets because of the lack of reliable and comparative data. There may also be cultural issues associated with the collection of data. It may be difficult to find experienced research resources in those countries to collect and analyse data. There may also be issues associated with translation.

These problems may be overcome by a number of ways, such as:

- Using secondary sources through reliable Government and trade organisations in the domestic market and through Embassies, Chambers of Commerce and so on

- Visiting the market(s) in question to gain a first-hand understanding of the market and its condition, perhaps through an organised trade mission

- Identifying experts that understand the local market conditions

Nevertheless you should accept a degree of error in your findings from emerging markets and treat them carefully.

5 The answer to this will largely depend upon your organisation. Such a system will provide managers with real-time market information, as customers worldwide become ever more knowledgeable and sophisticated. It is increasingly important that senior decision makers and global marketers be closely involved in the in the fieldwork of seeing and hearing the voice of the global consumer. It is worth identifying the extent to which there is such a system and ways in which it can be improved.

Cateora, P. and Graham, J., (2009). *International Marketing*. 14th edition. London: McGraw Hill.

Doole, I. and Lowe, R., (2008). *International Marketing Strategy Analysis, Development and Implementation*. 5th edition. London: Thomson Learning.

Hollensen, S., (2007). *Global Marketing*. 4th edition. London: Prentice Hall.

Schmidt, M and Hollensen, S., (2006). *Marketing Research – An International Approach*. London: FT/Prentice Hall.

Sethi, P., (1971). Comparative cluster Analysis for World Markets. *Journal of Marketing Research*, vol v111, pp 348-54.

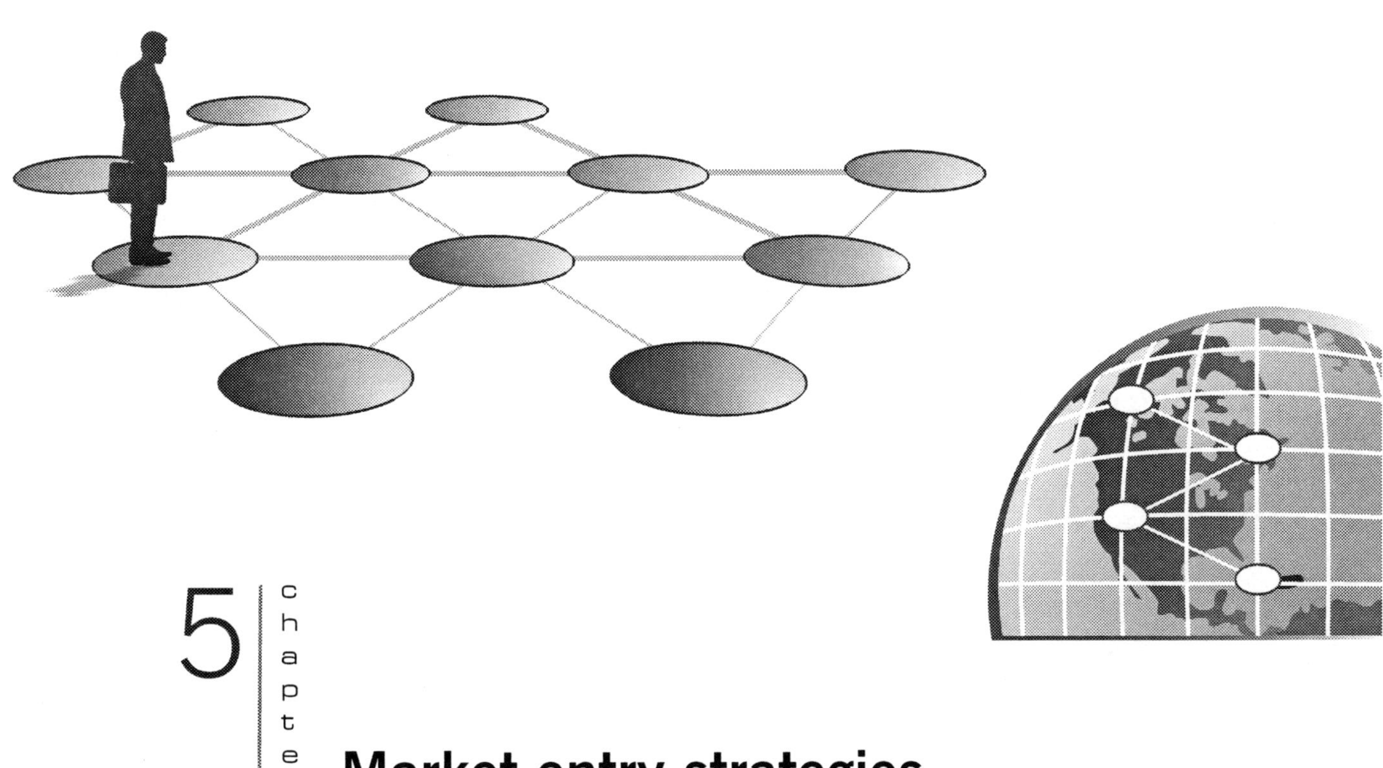

5 chapter

Market entry strategies

Once the organisation has chosen which markets to target it has to consider the best way to enter those markets. In this chapter we explore the major market entry modes and the criteria for selecting them. The entry modes they select will effect every area of their business model. For most small and medium sized enterprises this is a critical step. The challenge for established organisations is how to exploit the opportunities in each country more effectively. There is no ideal market entry strategy and different entry methods are likely to be adopted by different organisations entering the same market. An organisation may also choose to select a different market entry strategy for each country it enters. The chosen market entry mode is regarded as a critical element. In this session we explore the three main classifications of entry mode. There are advantages and disadvantages associated with each type of entry mode, each with differing degrees of risk, control and flexibility.

Contents

Chapter learning outcomes

In this chapter you will cover the following:

- Evaluate the entry mode selection strategy
- Consider alternative export modes
- Consider intermediate entry modes
- Consider hierarchical entry modes

1 Entry mode selection strategy

1.1 Three rules for entry mode selection

> **Definition**
>
> An **entry mode** is an institutional arrangement necessary for the entry of a company's products and services into a new foreign market. The main types are: export, intermediate and hierarchical modes.

There are three different rules (Root, 1994) that can be applied by an organisation in selecting its market entry strategy:

Naïve rule	The decision maker uses the same entry mode for all markets. However this rule ignores the differences of the individual foreign markets.
Pragmatic rule	The decision maker uses a workable entry strategy mode for each market, starting with a low risk entry mode. Only if the particular initial entry mode is not feasible or profitable will the organisation look for another workable entry mode. The workable entry mode is not necessarily the 'best' entry mode.
Strategy rules	The decision maker compares and evaluates all entry modes before a decision is taken. The may be determined against agreed criteria such as profitability, return on investment, speed to market, meeting customer needs and so on.

1.2 Factors influencing choice of entry mode

Generally speaking the choice of entry mode should be based on the expected contribution to profit. However this is easier said than done, particularly where market data is lacking, limited or unreliable.

Four factors are generally believed to influence the entry mode decision (Hollensen 2007):

- Internal factors
- External factors
- Desired-mode characteristics
- Transaction-specific behaviour

Internal factors	Comments
Organisation size	• Increasing resource availability provides the basis for increased global activity • Smaller enterprises may be forced to enter global markets without the right degree of control they ultimately desire

Internal factors	Comments
International experience	• Reduces cost and uncertainty of entering or supporting a given market
	• Increases profitability of an organisation committing resources to global markets
Product/Service	• Nature of the product will effect exporting mode and channel selection

External factors	Comments
Socio-cultural distance between home country and host country	• Socio-cultural differences create uncertainty for the organisation, thus influencing the mode of entry strategy
	• The greater the perceived distance in terms of culture, business practices and legal and economic systems the more likely the organisation will shy away from direct investment in favour of low resource commitments and high flexibility
Country risk/demand uncertainty	• Risk analysis should be undertaken to reduce potential for failure and opportunity for success
	• Where risk (ie political, economic) is high, organisations will favour entry modes where resource commitments are low
Market size and growth	• The larger the market, the larger the growth potential, the more likely management will commit resource to its development and seek to establish market entry modes which ensure greater control (ie establishing wholly-owned subsidiary)
	• The smaller the market the more an organisation may favour exporting or a licensing agreement
Direct and indirect trade barriers	• Tariffs or quotas favour the establishment of local production or operations.
	• Preferences for local suppliers, local knowledge, foreign language barriers or 'buy national' policies encourage an organisation to enter into a joint venture with a local organisation
	• Other barriers include trade regulations, customs formalities, product regulations and standards
Intensity of competition	• The greater the intensity of local competition the more the organisation will favour entry modes that involve low resource commitments (ie export modes)
Small number of relevant intermediaries available	• This will favour the requirement to favour an entry mode where more control is exerted in the market

Desired mode characteristics	Comments
Risk averse	• Risk averse decision makers will prefer export modes or licensing as they involve low levels of financial and management resource commitment
	• These are unlikely to foster development of a strong global strategy any may result in a significant loss of opportunity
Control	• Often linked to level of desired resource commitment
	• Modes of entry with minimal resource commitment provide little or no control over which the product or service is sold and marketed
	• Modes of entry such as wholly-owned subsidiaries provide most control but required the greatest commitment of resources

Desired mode characteristics	Comments
Flexibility	• Management must weigh up the flexibility associated with a given mode of entry

Transaction – specific factors	Comments
Tacit nature or know how	• If the nature of the organisation is difficult to articulate and express in words, because the product or service is complex, the more the organisation will involve a mode of entry where a high degree of control and commitment is exerted

Source: Table adapted based on Hollensen (2007)

The figure below (adapted from Doole, Lowe and Phillips, 1995) shows the balance between control and risk of the key market entry methods. Clearly, the greater the level of involvement, the greater the risk. Each of these entry modes will be covered in the subsequent parts of this section.

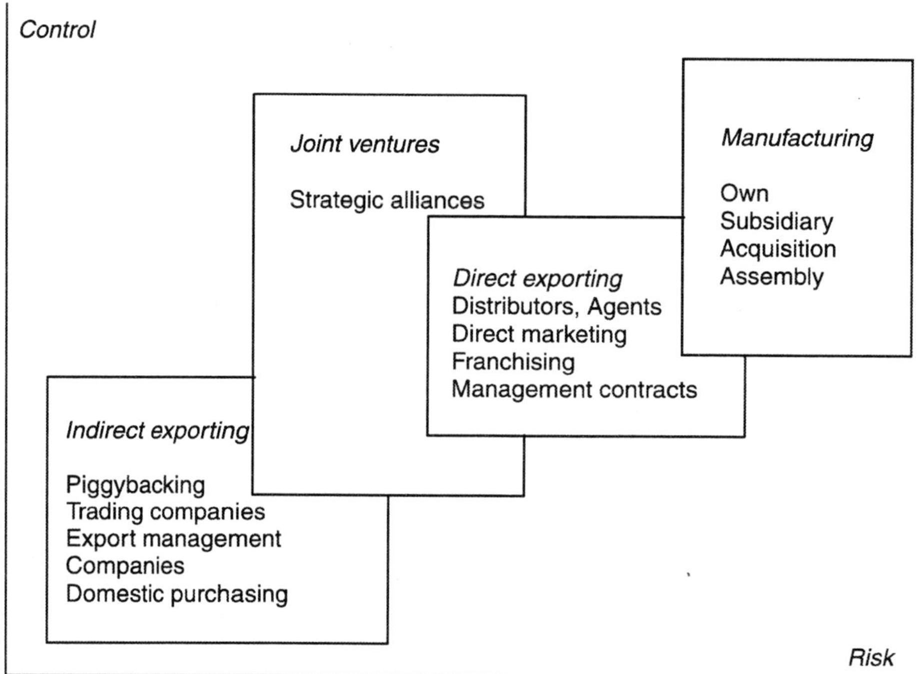

Selecting the market entry strategy is the key decision many companies have to take when expanding into foreign markets. Consider the key issues that your organisation may need to consider in making its decision on the choice of market entry method.

Hollensen (2007) has identified three broad categories of market entry modes and its is these we refer to in this Chapter.

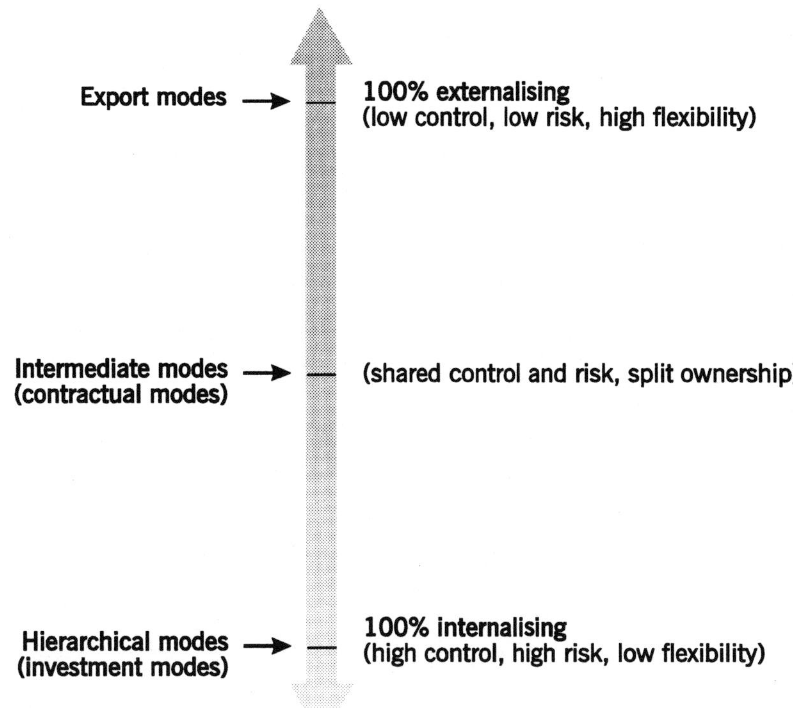

Adapted from Hollensen (2007).

2 Export modes

2.1 Exporting – an overview

Exporting is the easiest, cheapest and most commonly used route into a new foreign market. Many organisations become exporters in an unplanned, haphazard and reactive way, simply by accepting orders from potential customers who happen to be based overseas.

However, it is also common for an organisation to take a proactive approach to exporting, by systematic planning and the identification and selection of target markets for its exports. This gives rise to several advantages over those entry methods which require greater involvement in the overseas market.

- Exporters are able to concentrate production in a **single location**, giving economies of scale and consistency of product quality

- Organisations lacking the **know-how and experience** can try global marketing on a small scale

- Exporting enables organisations to **develop and test** their plans and strategies

- Exporting enables organisations to minimise their **operating costs**, administrative overheads and personnel requirements

Although exporting requires a **low involvement** in the overseas market, this does not necessarily imply that only low investment is needed. Exporting requires investment in **market research**, **strategy** formulation and careful **implementation of the marketing mix**. The initial success of Japanese car firms in the USA and Europe was based on research and strategic planning that was both extensive and costly.

2.2 Indirect export modes

Definition

Indirect exporting is where an organisation's goods are sold abroad by other organisations.

There are four main forms of indirect export.

- Export buying agent
- Export broker
- Trading company
- Piggyback or complementary exporting

2.2.1 Export buying agent

Export buying agents are organisations which facilitate exporting on behalf of the producer. There are three main types of export buying agent:

- **Export merchants** act as export principals. They buy goods from a producer and sell them abroad.

- **Confirming houses** also act as principals. Their main function is to provide credit to customers when the producer is unwilling to do so.

- **Manufacturers' export agents** are based at home, but sell abroad for the producer. An agent will usually cover a particular sector or industry, (eg pottery). Remuneration is by commission.

Advantages	Disadvantages
• The producer gains the benefits of the export house's market knowledge and contacts	• Ultimately, it is not the producer's but the merchant's decision to market a product, and so a producer is at the merchant's mercy.

• Except in the case of export agents the producer is relieved of the need to do the following
  – **Finance** the export transaction
  – Suffer the **credit risk**
  – Prepare **export documentation**
	• Any goodwill created in the market usually benefits the merchant and not the producer.
	
	• As with all intermediaries, an export house or merchant might service a variety of producing organisations. An individual producer cannot rely on the merchant's exclusive loyalty.

• The producer does not bear the **overhead costs** of export marketing.	• Export houses are not normally willing to enter into long term arrangements with a producer.
• In some cases export merchants receive preferential treatment from foreign institutional and organisational customers.	
• Where export agents are used, the producer retains considerable **control** over the market	

2.2.2 Export broker

Export brokers offer a full export management service. In effect, they perform the same functions as an in-house export department but are normally remunerated by way of commission.

Advantages	Disadvantages
• The manufacturer (or service provider) immediately gains its own export department without incurring overheads, retains full market control and can normally expect a long term relationship with the export broker	• As the export broker is an independent organisation, it can leave the producer's service and the producer will have gained **no in-house exporting expertise**
• Broker is typically a specialist in a given sector and can provide understanding and knowledge of the market and existing contacts	• As the producer does not learn from the experience of exporting, this may adversely affect future options by restricting those available. • The broker may not have sufficient knowledge of all the producer's markets

Developing Global Marketing Strategy

2.2.3 Trading company

Many foreign governments and foreign companies (eg department stores) have buying offices set up permanently in other countries. These go back to colonial days but are still important in the Far East and Africa.

2.2.4 Complementary exporting

Definition

Complementary exporting occurs when one producing organisation (the **carrier**) uses its own established channels to market the products of another producer (the **rider**) as well as its own (hence the term piggybacking is sometimes used). The carrier may act as:

- A simple transporter, using spare capacity in its distribution network
- An agent, selling the rider's goods for commission
- A merchant, buying and selling the rider's goods

It is often seen as a low-risk way for an organisation to begin expert operations and is especially well suited to manufacturers that are too small to invest directly or do not want to invest heavily in global marketing.

Carrier advantages	Carrier disadvantages
<ul><li>An organisation can get quick access to the product at low cost as no need for investment in R&D, production facilities or market testing</li><li>The 'rider' may be a good candidate for acquisition or joint venture leading to a stronger relationship</li></ul>	<ul><li>Some concerns will still exist about quality control and warranty</li><li>If a substantial market is built can the 'rider' keep with demand?</li></ul>

Rider advantages	Carrier disadvantages
<ul><li>Can export relatively conveniently without having to establish own distribution networks</li><li>Can learn from the 'carrier's' experience</li></ul>	<ul><li>Loses control over the marketing of its products</li><li>Lack of commitment may lead to loss of potential sales</li></ul>

Turnkey contracts may also provide opportunities for complementary exporting. A single firm engaged on a particular project overseas (eg construction and civil engineering projects in the Middle East) will often acquire products and services from other firms in the home country for the project.

2.3 Direct export modes

Definition

Direct exporting occurs where the producing organisation itself performs the export tasks rather than using an intermediary. Sales are made directly to customers overseas who may be the wholesalers, retailers or final users. Sales may increasingly be made via e-commerce on the Internet.

As exporters grow more confident they may decide to undertake their own exporting task. This will involve building up overseas contacts, undertaking marketing research, handling documentation and transportation, designing marketing strategies and so on.

There are two main forms of direct export:

- Distributors
- Agents

2.3.1 Agents

Strictly speaking an overseas export agent is an overseas firm hired to effect a sales contract between the principal (ie the exporter) and a customer.

Agents do not take title to goods; they earn a commission. In practice, however, the phrase is often understood to include distributors (who do take title). Some agents merely arrange sales; others hold stocks and/or carry out servicing on the principal's behalf.

As with all intermediaries, the use of an agent requires careful planning, selection, motivation and control.

Advantages of agents	Carrier disadvantages
- They have extensive knowledge and experience of the overseas market and the customers.	- An intermediary's commitment and motivation may be weaker than the producer's.
- Their existing product range is usually complementary to the exporter's. This may help the exporter penetrate the overseas market.	- Agents usually want steady turnover. Using an agent may not be the most appropriate way of selling low volume, high value goods with unsteady patterns of demand, or where sales are infrequent.
- The exporter does not have to make a large investment outlay.	- Many agents are too small to exploit a major market to its full extent. Many serve only limited geographical segments.
- The political risk is low	- As a market grows large it becomes less efficient to use an agent. A branch office or subsidiary company will achieve economies of scale.

Global case study

A French childrens clothing manufacturer wanted to break into the UK, Italian and Dutch markets. Although it was possible to use their French salesforce in Holland and Italy, it was not logistically possible to use the same team in the UK. Neither was it an option to set up a UK office so agents who specialised in sales to the top end of the market were used. The agents handled sales amongst premium retailers for a number of comeptitor brands. Although this might be considered a riskier option the company was able to enter a market it could not internally resource.

2.3.2 Distributors

Distributors are customers with preferential rights to buy and sell a range of a firm's goods in a specific geographical area. Distributors earn profit, not commission. Their selling and marketing activities on behalf of a producer are typically restricted geographically.

Stockists are distributors who receive more favourable financial rewards than distributors as they normally undertake to carry at least a certain minimum level of stock. The advantages and disadvantages of distributors and stockists are similar to those associated with overseas agents.

Developing Global Marketing Strategy

2.3.3 Choice of an intermediary

It can be difficult finding the right intermediary. The following sources can be helpful:

- Potential customers

- Recommendations from trade associations, chambers of commerce and government trade departments

- Word of mouth, conferences, trade magazines and networking

In order to assess one potential intermediary against another you could undertake a screening process that considers the following:

Criteria	Score (1-5)
How financially sound is the partner?	
How effectively does the partner cover the geographic area?	
How effective is the partner's marketing effectiveness and selling expertise?	
How competent is the partner's technical expertise?	
What is the reputation of the partner's level of service support?	
How well established is the partner's business networks?	
Does the partner have a potential conflict of interest with a direct competitor?	
TOTAL (out of 35)	

The success of a relationship with an intermediary can be put down to the following factors:

- Allocating time and resources to selecting the most suitable intermediary

- Ensuring that the expectations of each party are properly defined, understood and agreed

- Ensuring that the intermediary is suitably motivated to deliver and improve performance

- Providing adequate support on a continuing basis including training, joint promotion and developing contacts

- Ensuring that there is sufficient advice and information to/from both parties

2.4 Co-operative export modes/marketing groups

Export marketing groups are frequently found among small and medium sized enterprises attempting to enter the export market for the first time.

This enables small, specialist manufacturers to work together to form a broader concept that could be more attractive to a distributor, especially if the total product concept targets a certain lifestyle of the end customer.

In a loose arrangement this might involve manufacturers selling their own brands through the same agent. A tighter arrangement might result in the creation of a new export association. Characteristics of such an association include:

- Exporting in the name of the association
- Consolidating freight, negotiating rates and chartering of ships
- Performing market research
- Appointing selling agent abroad
- Setting process for export
- Creating uniform contracts for terms of sale
- Sharing marketing costs and undertaking joint campaigns

The CzechTrade Agency provides individual assistance to early-stage Czech exporters as well as to experienced ones to succeed in foreign markets. Services for Czech companies represent a wide range of activities. CzechTrade can also help foreign companies find suitable business partners from the Czech Republic through its headquarters in Prague and 33 foreign offices in more than 30 countries worldwide.

It delivers free and confidential services, which represent a reliable and time-saving way of building business relationships. The assistance among others involves supplier search, business visits assistance, organising business meetings, publication of foreign offers or assistance of our foreign representatives.

http://alliances.czechtrade.cz/en/

Advantages	Disadvantages
• Organisations can reach foreign markets more effectively together, with better representation	• Loss of independence of the founders of the respective organisations
• Create more stable prices and can reduce selling costs	
• Creation of a stronger brand name	

Why is exporting frequently considered the simplest way of entering a foreign market and is this favoured by small and medium sized enterprises? What are the disadvantages of adopting such a market entry mode?

3 Intermediate entry modes

3.1 Intermediate entry modes – an overview

Definition

Intermediate entry modes are primarily used for the transfer of knowledge and skills, although they may also create their own export opportunities. Ownership and control is shared between the parent firm and a local partner.

The most common form of intermediate entry modes are:

- Contract manufacturing
- Licensing
- Franchising
- Joint ventures and strategic alliances

Each of these is now discussed in detail.

> **Definition**
>
> A **licensing agreement** is a commercial contract whereby the licenser gives something of value to the licensee in exchange for certain performances and payments.

The licenser may provide any of the following.

- Rights to produce a patented product or use a patented production process
- Manufacturing know-how (unpatented)
- Technical advice and assistance
- Marketing advice and assistance
- Rights to use a trademark, brand etc.

Advantages	Disadvantages
- It requires no investment, save the continuing costs of monitoring the agreement. - It enables entry into markets that would otherwise be closed (eg by tariffs, government attitude and policies) - As a method of entry, it is relatively simple and quick - The licenser gains access to knowledge of local conditions - New products can be introduced to many countries quickly because of low investment requirements - It provides all the usual benefits of overseas production (cheaper transport, lower import barriers and so on) - It can be a source of competitive advantage, if it spreads the firm's proprietary technology, giving it wider exposure than that of a rival	- Revenues from licenses are very low, usually less than 10% of turnover. The significance of this naturally depends on the profit margins that might otherwise be expected - A licensee may eventually become the licenser's competitor. During the license period, the licensee may gain enough know-how from the licensor to be able to operate independently - Although the contract may specify a minimum sales volume, there is some danger that the licensee will not fully exploit the market - Product quality might deteriorate if the licensee has a more lax attitude to quality control than the licenser - Governments may impose restrictions or conditions on the payment of royalties to the licenser or on the supply of components - It is often difficult to control the licensee effectively. The licensee's objectives often conflict with those of the licenser and disagreements are common

Astute management of the license agreement is essential.

- Be careful in the choice of licensees (eg identify criteria that must be satisfied)
- Design contracts that protect both parties
- Control the licensee, by, say, having an equity interest in the licensee's business or by retaining control over key input components
- Take action to motivate the licensee

Activity 3

Do you believe that licensing represents a feasible market entry strategy for your organisation? Provide examples of organisations that have adopted a licensing approach.

> **Definition**
>
> A **franchise agreement** specifies, in more detail than a license agreement, exactly what is expected of the franchisee. In a franchise arrangement, the franchiser supplies a standard package of goods, components or ingredients along with management and marketing services or advice. The franchisee supplies capital, personal involvement and local market knowledge. Avis, Holiday Inn, Pepsi Cola, Kentucky Fried Chicken, the Body Shop, and McDonald's have franchise arrangements in many countries.

A number of factors have contributed to the rapid growth of franchising:

1   Decline of traditional manufacturing industry and its replacement by service-sector industries
2   Growth in the popularity of self-employment and business ownership

There are generally two type of franchising:

- **Product and trade name franchising**

    Similar to trade mark licensing. Typically it is a distribution system in which suppliers make contracts with dealers to buy or sell products or product lines. Dealers use the trade name, trademark and product line.

- **Business format package franchising**

    This involves a relationship between the entrant (franchisor) and host country entity (franchisee) in which the former transfers, under contract, a business package (or format) that it has developed and owns, to the latter.

    The package generally contains the following items:

    - Trademarks/names
    - Copyright
    - Designs
    - Patents
    - Trade secrets
    - Business know-how
    - Geographic exclusivity
    - Location selection

The franchise system combines the advantages of economy of scale with the local knowledge and entrepreneurial talents of the franchisee. Their joint contribution may result in its success.

The franchisor depends on franchisees for fast growth, an infusion of capital from the franchise purchase fee, and an income stream from the royalty fee paid by franchisees each year. The franchisor also benefits from franchisee goodwill and motivation to operate a successful independent business.

The franchisee relies on the franchisor for the strength of the trademark, technical advice, support services, marketing resources and promotional support.

A possible downside of this market entry method is the failure of either party to live up to the terms of the legal agreement. Conflict and disagreement can be reduced by each party viewing the other as a partner in the success of the relationship. This requires a strong common culture with shared values established through intensive communication.

Global case study

McDonald's is an example of brand franchising. McDonald's, the franchisor, grants the right to sell McDonald's branded goods to someone wishing to set up their own business, the franchisee. Under a McDonald's franchise, McDonald's owns and leases the site and the restaurant building. The franchisee

buys the fittings, the equipment and the right to operate the franchise for 20 years. To ensure uniformity throughout the world, all franchisees must use standardised McDonald's branding, menus, design layouts and administration systems. The licence agreement also insists the franchisee uses the same manufacturing or operating methods and maintains the quality of the menu items. McDonald's recognises the benefits of a franchised operation. Franchises bring entrepreneurs full of determination and ideas to the organisation. Franchising enables McDonald's to enjoy considerably faster growth and the creation of a truly global brand identity. The more restaurants there are, the more McDonald's can benefit from economies of scale. On the financial side, McDonald's receives a monthly rent which is calculated on a sliding scale based on the restaurant's sales, i.e. the higher the sales, the higher the percentage and vice versa. There is also a service fee of 5% of sales that is contributed to support department activities and royalties. The purchase price of a restaurant is generally about £150,000 upwards. The new franchisee is expected to fund a minimum of 25% of this from their own unencumbered funds.

http://www.mcdonalds.co.uk/static/pdf/aboutus/education/mcd_franchising.pdf (2008)

3.1.3 Joint ventures and strategic alliances

Definition

A **joint venture** is an arrangement where two or more (often competing) firms join forces for manufacturing, financial and marketing purposes and each has a share in both the equity and the management of the business, sharing profits, risks and assets.

Forming a joint venture with a technologically advanced foreign company can lead to new product development, maybe at a lower cost. Licensing, franchising and contract manufacture are loose forms of joint venture.

However joint ventures are bound by much stronger formal ties. They essentially focus on a single national market. When based abroad, they usually involve partners of unequal strength, for example when a developed country multinational, contributing capital and technology joins forces with a local firm in a developing country, offering local market knowledge and contacts.

Advantages	Disadvantages
• As the capital outlay is shared, joint ventures are attractive to smaller or risk-averse firms, or where very expensive new technologies are being researched and developed (such as the civil aerospace industry).	• There can be major conflicts of interest between the different parties in terms of aims, objectives and overall strategy
• When funds are limited, joint ventures permit coverage of a larger number of countries since each one requires less investment by each participator.	• Substantial commitment of investment of capital and management resources required to ensure success, particularly in terms of building trust and understanding each others needs and concerns
• A joint venture can reduce the risk of government intervention as a local firm is involved (eg Club Med pays much attention to this factor). The strong role of government in China means that nearly all foreign ventures in China are alliances with Chinese partners.	• There is often a problem with managing the different cultures between both organisations. Also ignoring local cultural issues will destroy the opportunity for success
	• Some protectionist governments discourage or even prohibit foreign firms setting up independent operations. Instead, they encourage joint ventures with indigenous firms

Advantages	Disadvantages
• Licensing and franchising often give a company income based on turnover, and any profits from cost reductions accrue to the licensee. In a joint venture, the participating enterprises benefit from all sources of profit.	• Some governments regard uncontrolled investments from overseas as a type of colonial exploitation and are averse to sending foreign exchange outside the country
• Joint ventures can provide close control over marketing and other operations.	
• A joint venture with an indigenous firm provides local knowledge. This is a big advantage for firms seeking to do business in difficult markets, such as Russia. Political know-how, site selection expertise and business connections are important.	
• In oligopolistic markets, where a few firms are dominant, a foreign firm may find the cost of market entry too high, and seek an alliance with an established competitor	
• Joint ventures generally involve a transfer of know-how and technology that benefits the local economy. Joint venturing between outside and local firms is encouraged, for example, in India, Nigeria and the former USSR	

The Seven Steps in establishing a successful joint venture

Step 1:   Establish and agree clear objectives of the joint venture and specify the time period for its achievement

Step 2:   Evaluate the advantages and disadvantages of establishing a joint venture compared with alternative market entry strategies for achieving the objectives

Step 3:   Identify, screen, meet and select a joint-venture partners against a set criteria

Step 4:   Develop the business plan and agree broad agreement on the various business issues

Step 5:   Negotiate the joint-venture agreement

Step 6:   Write and agree the contractual relationship

Step 7:   Evaluate the performance of the joint venture at regular intervals and against agreed performance measures

Around 80% of joint ventures fail for one reason or another. Given the high failure rate both parties should prepare for such an eventuality by addressing the issue in the contact.

> ## Definition
>
> **Strategic alliances**
>
> At least two organisations combining value chain activities for the purpose of competitive advantage (Bronder and Pritzl, 1992).

The participants tend to be competing firms from different countries, seeking to enhance their competencies by combining resources, but without sacrificing autonomy. The strategic alliance is usually concerned with gaining market entry, remaining globally competitive and attaining economies of scale. They can be broadly categorised as follows.

- **Production** based alliances - improving manufacturing and production efficiency
- **Distribution** based alliances - sharing distribution networks
- **Technology** based alliances pooling R & D costs

Alliances may be horizontal (between two firms in the same industry) or vertical, involving collaboration between a supplier and a buyer. Sometimes, they involve firms with no such connection.

Strategic alliances have mainly concentrated in manufacturing and hi-tech industries, and increasingly in services (notably airlines, such as the 'One World' alliance involving British Airways, Qantas and several other international collaborators).

A five-point checklist of 'Cs' has been suggested for choosing an alliance partner:

- Complementary skill sets and products
- Capability to enter new markets
- Clear understanding of the commercial arrangements
- Chemistry between the participants
- Shared intellectual capital

The reasons that drive the formation and operation of strategic alliances can be identified as follows:

Driving forces	Comment
Insufficient resources	Because of finite resources powerful and independent organisations need to co-operate in order to remain competitive
Pace of innovation and market diffusion	Strategic alliances can succeed in combining resources to ensure the value chain is protected and enhanced in the face of competition from other players
High research and development costs	Co-operation can reduce the costs of R&D and increase value to the end customer
Concentration of organisations in mature industries	Many industries have used alliances to manage the problem of excess production capacity in mature markets
Self protection	Alliances are formed to protect against stiff competition
Market access	Strategic alliances can be used to gain access to difficult markets

The success of strategic alliances depends on effective management, good planning, adequate research, accountability and monitoring. As with joint ventures it is important to recognise potential cultural sensitivities in cross-border alliances and the implications this has for trust and quality information exchange.

Starbucks opened a branch in the Forbidden City in Beijing, China, on September 18, 2000. Controversy on the Internet began shortly thereafter. A poll by the popular portal *Sina.com* showed that more than 70% of nearly 60,000 people surveyed were opposed to Starbucks' entry into the Forbidden City, the main reason being the damaging effect on a Chinese cultural heritage and its atmosphere, according to a report by the People's Daily, the Communist Party's mouthpiece, on November 24, 2000. Starbucks was soon forced to remove its trademark green sign in favour of the Chinese characters for Xing Ba Ke, which now adorns the window. The Starbucks controversy also led to public discussion about the KFC chain that was to be withdrawn from Beihai Park northwest of the Forbidden City.

http://www.atimes.com/atimes/China_Business/IA25Cb02.html, 25 January 2007

Study buddy

Share your thoughts with your Study buddy as to the advantages and disadvantages of establishing joint ventures and strategic alliances between organisations from each of your countries.

4 Hierarchical entry modes

Establishing and running a production facility in an overseas market demonstrates the fullest commitment to that market. Production capacity can be built from scratch, or an existing organisation can be acquired. The main reasons for investing in the creation of a local operation include:

- Gaining new business
- Defending existing business
- Moving with an established customer
- Saving costs
- Avoiding government restrictions

4.1 Hierarchical entry modes – an overview

The main hierarchical mode of entry is often referred to as a **wholly owned subsidiary.**

Advantages	Disadvantages
• The organisation does not have to share its profits with partners of any kind	• The substantial investment funding required prevents some firms from establishing operations overseas
• The organisation does not have to share or delegate decision making and so there are no losses in efficiency arising from inter-firm conflict	• Suitable managers, whether recruited in the overseas market or posted abroad from home, may be difficult to obtain
• There are none of the communication problems that arise in joint ventures, license agreements etc.	• Some overseas governments discourage 100% ownership of an enterprise by a foreign company
• The organisation is able to operate a completely integrated and synergistic international system	• This method of entry forgoes the benefits of an overseas partner's market knowledge, distribution system and other local expertise
• The organisation gains more experience from overseas production	

Developing Global Marketing Strategy

Establishing a wholly owned subsidiary can be undertaken by acquisition or through organic growth (otherwise known as a greenfield investment).

Acquisition, as a method of entry, is rapid and offers the benefits of an existing management team, market knowledge and all the other trappings of a 'going concern'.

Acquiring a company can also be a way of ensuring a market presence in an overseas market or segment. Acquiring an overseas subsidiary has the following justifications, in marketing terms.

- Shared distribution networks (ie getting more share out of your existing network)
- Access to new markets, in which the acquired company already has a presence
- Access to new brands, so that a variety of brands can be promoted

Organic growth offers fewer of the 'quick' advantages of acquisitions: it involves building up a presence from scratch. The disadvantages are the time and effort it takes to build a new market presence, especially in mature markets, where growth in market share involves a war of attrition, and where a firm has little knowledge of the market. However, it offers control over the process of growth, and there need be fewer clashes of corporate culture. For new products it might be better than acquisitions.

Activity 4

Consider why acquisition is often the preferred way to establish a wholly-owned subsidiary abroad? What are the limitations of acquisition as an entry method?

Assessment advice

You will be expected to be able to evaluate critically appropriate market entry strategies for organisations considering global expansion, and their impact on the organisations existing strategy, resources, processes and systems. This session should enable you to consider the wide range of market entry modes available for your organisation and enable you to consider those that are most appropriate given your organisation. Collect examples of organisations from around the world that reflect the different modes of entry, determine why they are following a certain strategy, what the alternative options are they could have considered and the lessons you can learn from their experiences.

4.2 Withdrawing from an overseas market

Closing down a wholly owned subsidiary or selling it to another organisation is a strategic decision that must be taken if profits are too low, demand is decreasing or competitors are more efficient for example.

Benito (1996) identifies four main factors that will have an impact on the likelihood of divestment:

Factor	Comments
Environmental stability	• Investments made in R&D and marketing
	• Political risks may lead to forced divestment
Attractiveness of current operations	• Unsatisfactory performance
	• If it is performing well the owner may feel they can get a good price
	• Economic growth increases barriers to exit
	• Attractiveness of location makes the operation a more likely takeover target
Strategic fit	• Unrelated expansion or diversification increases cost of governance

Factor	Comments
Governance issues	• Cultural distance – the closer the culture to the host the higher the barriers to exit and vice versa
	• Lack of parent company commitment will increase incentive to exit
	• As organisations gain experience in running an overseas subsidiary they will seek to find solutions rather than divest of a subsidiary

Global case study

Olympus Corporation Agrees to Strategic Divestment of Diagnostic Systems Business

Since Olympus entered the diagnostic systems market in 1971 almost four decades ago, Olympus has been engaged in the manufacture and marketing of clinical diagnostic testing systems, principally automated chemistry analyzers and automated blood transfusion testing systems. However, the presence of several large existing competitors, an increase in M&A activity and the entry in recent years of significant new players from other industries have created a new competitive environment in this market segment.

Taking into consideration the rapidly changing dynamics within this global market segment, Olympus determined that the divestment of its diagnostic systems business to Beckman, a major player in this segment, was the best strategic option and offered clear benefits going forward over retaining ownership and continuing to operate this business within the Olympus Group.

In addition to realising a fair value for this business, Olympus also carefully considered this divestment from personnel, organizational and corporate culture perspectives. Olympus was confident that this agreement would expand the number of opportunities and range of possibilities for both this business and its employees

http://www.olympus-global.com/en/news/2009a/nr090227dse.cfm, Accessed 31 March 2010

Developing Global Marketing Strategy

1 Evaluate the entry mode selection strategy

- The choice of market entry mode is regarded as a critical first step in the process of global marketing since it sets the pattern for future international engagement.

- There are a number of market entry methods that differ in risk, control, commitment and investment. An organisation must select the entry mode that is most appropriate to their size and type of organisation, type of products and services and the country it intends to enter.

2 Consider alternative export modes

- Each market entry mode has a number of advantages and disadvantages. It is your role as a global marketer to consider the options and provide a justification for the proposed market entry modes.

- Given the complex and changing nature of the global business environment increasingly complex and innovative operations are being created to balance the opposing forces of competitiveness and co-operation.

3 Consider intermediate entry modes

- Intermediate entry modes are primarily used for the transfer of knowledge or skill.

- A number of methods include contract manufacturing, licensing, franchising, joint ventures and strategic alliances.

4 Consider hierarchical entry modes

- Hierarchical modes of entry entail wholly owned subsidiaries.

- Achieved by acquisition or organic growth.

- Closing down a wholly owned subsidiary or selling it to another organisation is a strategic decision that must sometime be taken.

1   There are four types of factors that an organisation will have to take into consideration in deciding on the market entry strategy for overseas markets. The answer to this question will largely depend on your organisation. You can use the factors listed to evaluate the options available and develop a range of potential recommendations.

2   Exporting is the easiest, cheapest and most commonly used route into a new foreign market. Many organisations become exporters in an unplanned, haphazard and reactive way, simply by accepting orders from potential customers who happen to be based overseas.

- Exporters are able to concentrate production in a single location, giving economies of scale and consistency of product quality

- Organisations lacking the know-how and experience can try global marketing on a small scale

- Exporting enables organisations to develop and test their plans and strategies

- Exporting enables organisations to minimise their operating costs, administrative overheads and personnel requirements

Although exporting requires a low involvement in the overseas market, this does not necessarily imply that only low investment is needed. Exporting still requires investment in market research, strategy formulation and careful implementation of the marketing mix.

There are a number of forms of exporting. Each method should be considered and evaluated against specific selection criteria.

3   Whether or not licensing represents an opportunity depends to a large extent on the product characteristics, business strategy and business model of your organisation. There are a number of advantages and disadvantages associated with licensing that would need to be evaluated.

4   Acquisition, as a method of entry, is rapid and offers the benefits of an existing management team, market knowledge and all the other trappings of a 'going concern'. Acquiring a company can also be a way of ensuring a market presence in an overseas market or segment.

Establishing a wholly owned subsidiary offers fewer of the 'quick' advantages of acquisitions: it involves building up a presence from scratch. The disadvantages are the time and effort it takes to build a new market presence, especially in mature markets, where growth in market share involves a war of attrition, and where a firm has little knowledge of the market.

However, it does offer control over the process of growth, and there need be fewer clashes of corporate culture. For new products it might be a better option than an acquisition strategy.

Benito, G.R.E., (1996) 'Why are foreign subsidiaries divested? A conceptual framework' in I. Björkman and M. Forsgen, eds *The Nature of the International Firms*. Copenhagen: Copenhagen Business School

Bronder, C. and Pritel, R., (1992). 'Developing strategic alliances: a conceptual framework for successful co-operation'. *European management – Journal*, Vol 10 p412–421.

Cateora, P and Graham, J., (2009). *International Marketing*. 14th edition. London: McGraw Hill.

Doole, I and Lowe, R. and Phlips, E., (1994). *International Marketing Strategy, Analysis, Development & Implementation*. London: RO

Doole, I. and Lowe, R., (2008). *International Marketing Strategy Analysis, Development and Implementation*. 5th edition. London: Thomson Learning.

Hofstede, G., (1980). *Culture Consequences: International Differences in Work Related Values*. Sage Publishing.

Root F.R., (1994). *Entry Strategies for International markets*. Washington: Lexington Books.

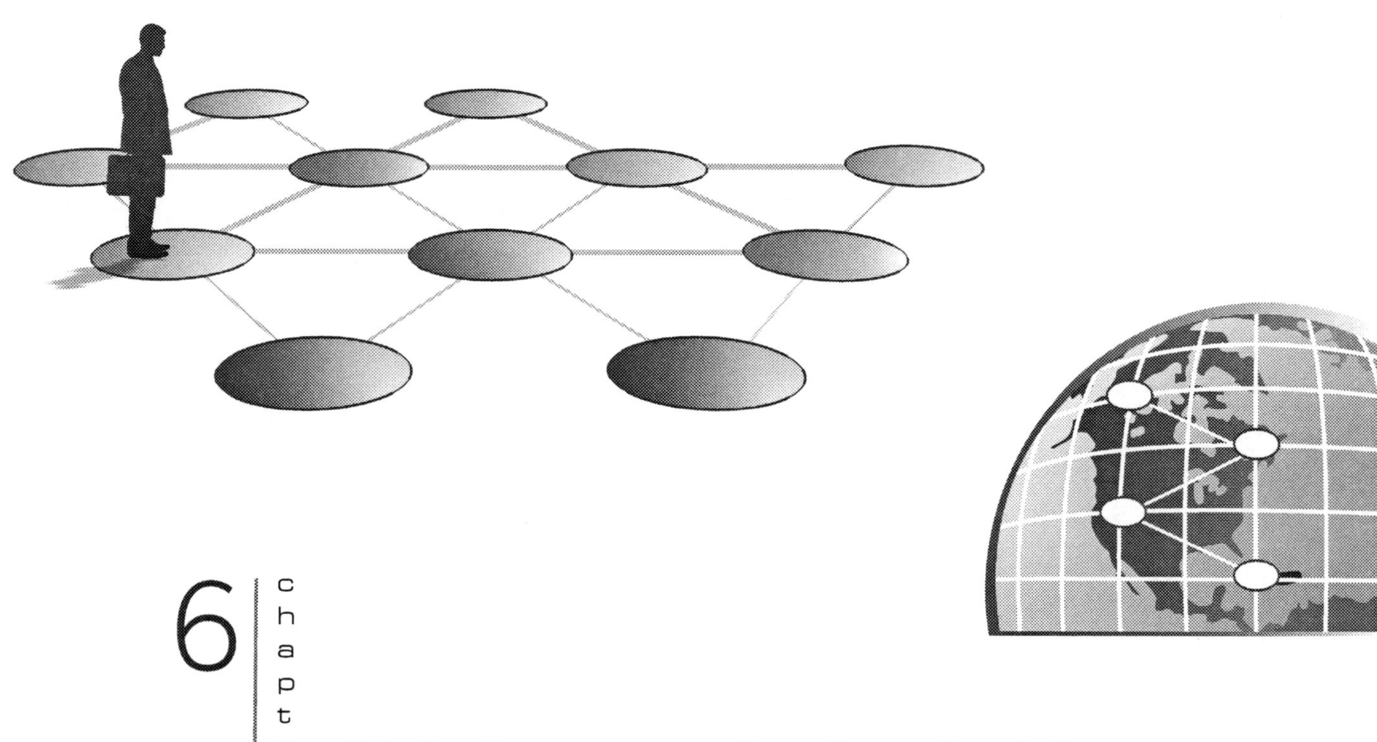

6 | chapter

Global communication decisions

Global communications are often the focus of many programmes referring to international marketing. This chapter begins by considering the issues impacting on communications on a global scale. The specific components of an effective communication plan are then covered.

These sections further the knowledge you gained in the Developing Integrated Communications Strategy module. The chapter concludes by considering how agencies should be selected when incorporating a global communication plan.

Contents

Chapter learning outcomes

In this chapter you will cover the following:

- Consider the issues in global marketing communications
- Understand the cultural considerations in global marketing communications
- Consider the legal aspects in global marketing communications
- Consider the media aspects in global marketing communications
- Understand the wider market considerations
- Evaluate how to select global agencies and manage them successfully

Assessment advice

It is now useful to evaluate the distribution strategy for the organisation you are investigating. Consider the options in light of current and past performance and evaluate the distribution strategy you are recommending for the country you are proposing to enter.

1 Setting the scene

Communications decisions are among the most sensitive considerations in global marketing. After all, marketing communications are the organisation's public face.

Not only does the legal environment differ (countries have different laws and regulations regarding advertisements) but the advertisements and other associated communications must be in keeping with the culture of the country, as well as the aspirations of the market segment within the country that the firm is selling to.

An important strategic consideration is whether to standardise globally or adapt the promotional mix to the environment of each country.

2 Issues in global marketing communications

Effective communications are particularly important in global marketing, due to the geographical and psychological separation of the producer and its market. Consequently, the global marketer has to rely more on intermediaries and impersonal communication for a major part of the communication process.

The methods of promotion in any one market will be affected by the following.

- The promotional objectives for that market
- Cultural constraints
- Legal constraints
- Facilities available for promotional effort in the market
- Economic development
- Distribution infrastructure
- Media availability
- Competition

Standardisation is rare and can only occur when the product and its cultural meaning is more or less identical across a number of countries and cultures. The more normal state of affairs is that of adaptation of promotional effort to match the different communication requirements of the market in question.

Developing Global Marketing Strategy

2.1 Promotional objectives

In new or developing markets the promotional objectives may well be to create awareness and encourage trial. In mature markets the objectives will be to encourage repeat purchase and remind the customer. The objectives chosen will thus influence the promotional campaign for a market. Thus in a new market the use of 'samples' may be used to encourage trial but in a mature market the use of 'price off next purchase' coupons (if allowed) might be more appropriate.

2.2 Culture

Our day to day lives are centred around how we communicate with each other and this is closely influenced by our cultural background. To succeed in a market which is not the company's home market requires a high level of understanding of the target market. Failure, more often than not, shows through in the form of mistakes, failures and omissions in the planning and control of the marketing communication programme.

Culture may be loosely defined as 'the way we do things around here' and is reflected in religion, schooling, aesthetics, peer groups, legal framework, language and so on. It is usually the area which provides the greatest opportunity for companies to make mistakes.

2.3 Planning

If an organisation is to avoid the many pitfalls awaiting the unwary in terms of marketing communications, the key is sound planning which will allow sound decisions to be made on the following issues:

Standardisation/ adaptation	Whether or not the product or service is standardised internationally, there will be a separate decision as to whether the communications should be standardised or adapted to local markets.
Push and pull strategies	Differing channel pressures and conditions may also lead to communication difficulties which need to be recognised and planned for.
	Organisations do not always get the planning correct, despite detailed marketing research and forethought.
Media availability	It is significant that different country markets have a different balance of media available for advertising support – so that, even if an organisation wished to have exactly the same communications strategy in each market, it would not necessarily always be possible.
Political and legal factors	May be different in each of the markets in which the company operates and this will provide very different ground rules.
Management	Multi-country marketing communications may require significantly different management approaches, depending upon the degree of standardisation of the message and the medium.
Centralisation	As with the standardisation issue there is a balance to be struck between pressures to centralise decision making in the communications field and the pressures to decentralise

The analgesic Panadol was seen by its brand management team to be able to benefit from a global brand recognition, using the UK packaging pattern and logo as the ideal. The UK packaging was blue and the logo was circular and easy to recognise. Different European markets had, for a variety of reasons, acquired different packaging – some packs were green and had a different logo and name. The Scandinavian market, different again, had a mainly white pack because customers expected medicines to appear in white packs. Other countries used the brand name Panodol, with yet another logo.

In the event the organisation standardised the logo shape (still using both names, however) and also co-ordinated the pack shape, leaving the different colours for continued recognition within each of the markets. Thus, some global branding was possible but different cultural background forces were catered for. The organisation had to hope that a reasonable balance was struck between the pressures for standardisation and the necessary local specialisation in the packaging.

2.4 Control

Planning should not be done without also putting into place the means whereby the marketing communications effort is effectively controlled. Control issues in global marketing invariably revolve around the following factors.

Objectives	Clear communications objectives must be established which provide a benchmark against which the success of an international communications campaign may be measured.
Standards	Clear standards must be set across international markets for the delivery of the communications message in a way that is appealing and satisfactory. This may be associated with media quality standards, or with creative campaign planning standards.
Management issues	Multi-country delivery of communication messages must be tightly controlled to achieve the best effect and to deliver the best value for money. This inevitably raises issues associated with the management of resources across international borders. Here again a balance must be managed between clarity of control and the benefits associated with local market knowledge.

A multinational organisation may also be using more than one advertising or promotions agency. A clear set of guidelines needs to be established to ensure accountable reporting lines between client and agencies.

Study buddy

Discuss with your Study buddy the key issues involved in planning, executing and controlling an integrated global communications strategy for your organisations. Then together share your thoughts with the wider cohort through the discussion forum.

3 Cultural considerations

Culture, as we have seen, is a term used to describe the set of values, beliefs, norms and artefacts held in common by a social group. Marketing communicators need to be particularly sensitive to culture in promotion if these messages are to work in global markets.

The following dimensions of culture are of particular relevance to the global marketing communicator:

- Verbal and non verbal communications
- Beliefs and values
- Aesthetics
- Family roles and relationships
- Learning
- Dress and appearance
- Work habits

3.1 Language and non verbal communication

Commonsense dictates that care must be taken when translating copy from one language to another. A literal translation of an advertising slogan would be meaningless in many countries. For example:

(a) The spirit of a communication message is preferred to word-for-word translation.

(b) Which language? It may be difficult to decide exactly which language to choose for translation purposes. In Canada, although the majority of the population speak English, packaging must include a French translation. Sometimes English is used to denote the global nature of the brand.

(c) Brand and product names must be assessed for their suitability in international markets. For instance Ford's Mondeo name works well in European markets, where the 'world' association comes across powerfully. In the US market however, the car is called the Contour.

A slogan or copy that is effective in one language may mean something completely different in another. Thus names, presentation materials and advertisements may well need to be adapted and translated when used in other markets.

The following global communication process highlights the key attributes of effective global communication (Hollensen 2007).

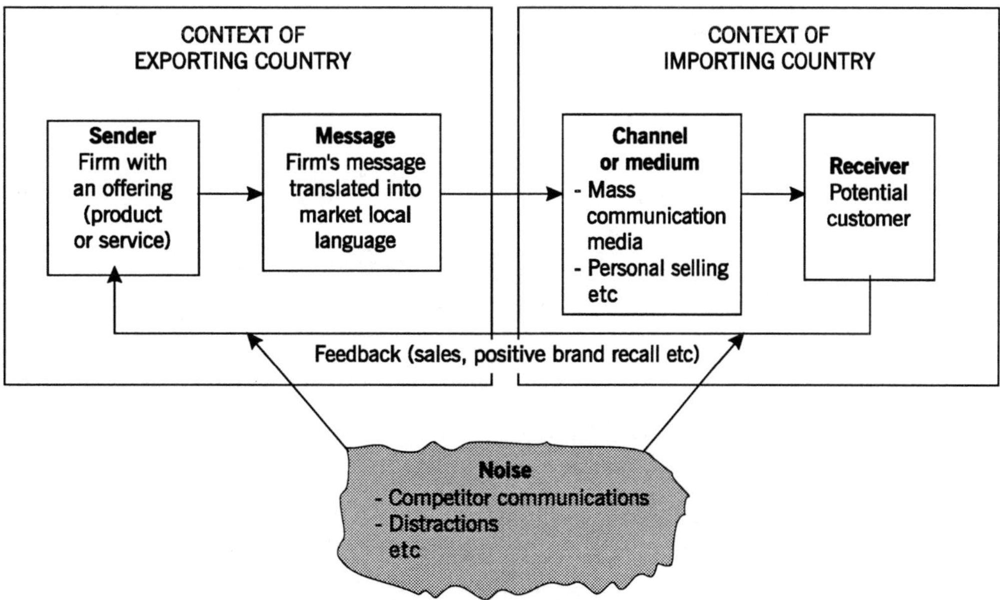

Adapted from Hollensen (2007)

3.2 Aesthetics

Attitudes towards different design and colour aesthetics vary around the world.

(a) **Symbols**. A fragrance or toiletries carton decorated with chrysanthemum flowers would not be a success in France, where the flowers are traditionally associated with funerals.

(b) **Visual representation**. The ideal of showing a figure partially out of frame is well known in Europe. In much of Africa by contrast, figures that go over the edge of their frame transgress the cultural rule about how pictures should look.

3.3 Dress and appearance

Advertisers need to be aware of cultural dress codes when deciding if an execution prepared in one market will be acceptable in another.

3.4 Family roles and relationships

Family has always been the dominant agent for transmitting culture.

(a) **Family structure**. In many third world countries, the extended family lives in a tight knit community. This contrasts with the nuclear family, high divorce rates etc, in the West.

(b) **Family roles** can differ greatly from country to country.

3.5 Beliefs and values

Beliefs and values evolve from religious teachings, family structures and the pattern and nature of economic development in that society.

3.6 Education and literacy

The level of education within a culture is an important factor for the global marketing communicator. Low literacy levels will mean that visual methods of communication take precedence over textual ones.

3.7 Work habits

Not all societies conform to the same work patterns, even though, with flexible employment patterns, this is fast disappearing.

3.8 Advertising culture

In addition to being sensitive to a country's culture in general, global marketers should be aware of the level of a country's advertising literacy. There can be five levels of advertising development along a continuum from the unsophisticated to the sophisticated:

1   Least sophisticated. The emphasis of advertising is on the manufacturer's description of the product.

2   At the next level, consumer choice is acknowledged so emphasis switches to the product's superiority over the competition.

3   Mid point. Consumer benefits are emphasised rather than product attributes. Executional devices may include the use of celebrity endorsements; role models may give demonstrations.

4   At a more sophisticated level, brands and their attributes are well known, so need only passing references (perhaps by way of a brief pack shot or logo).

5   At the most sophisticated level, the focus is sometimes on the advertising itself. The brand is referred to only obliquely, perhaps at a symbolic level, (eg Benson & Hedges). Consumers are believed to have a mature understanding of advertising, and are able to think laterally in order to decode messages.

Identify the key cultural issues that your organisation must face in developing a successful communications campaign for a given country. Compare and contrast these issues with the domestic market.

4 Legal considerations

Laws and regulations governing marketing communications must obviously be observed. Each country will have its own set of restrictions which apply to advertising, packaging, sales promotion or direct marketing.

Advertising	In some countries, restrictions apply to the use of non native models and actors. This can mean that advertising has to be reshot for specific countries.
Packaging regulations	Can vary. In a number of European markets, the push towards environmentally friendly packaging has resulted in far more stringent rules than apply in the UK. In Denmark, soft drinks may not be sold in cans, only in glass bottles with refundable deposits.
Direct marketing	There are laws requiring that consumers have the choice of opting out of receiving further mailings.
Promotional methods and claims	In the UK, for example, although considerable tightening up is in progress, the attitude towards promotional claims is relatively liberal, as is the attitude to what is promoted, when and where. Legal constraints may exist: • What can be claimed • What products may be promoted • What media may be used • When they may be promoted • How much may be spent on promotion
What can be advertised	Different countries in the EU, for example have a variety of restrictions on advertisements directed to children. Advertising to children under 12 is banned in Sweden. In the USA, such adverts are normally unlimited.
Data protection	There are a number of legal considerations to do with storing and sharing customer data, access to data in terms of freedom to view and the use of unsolicited mailings and contact.

5 Media considerations

Media is a complex, highly specialised area within marketing communications. You need to have a grasp of the factors influencing media choice in overseas markets, and some understanding of the practical problems which may be experienced when planning the use of media. Coverage by various media can be limited.

• Localised media organisations
• Circulation/audience limitations
• Legal restrictions

Media availability can vary greatly from country to country.

(a)   Some countries such as the UK offer a great variety of choice to the advertiser. Press is a well segmented sector with over 6,000 consumer, trade and technical magazines, as well as the national and regional newspaper titles. Commercial television includes terrestrial and satellite channels. Radio as a sector is strengthening, with both local and national stations. Cinema and outdoor are mature media.

(b)   In other countries, not all media are available, and the global marketer will have to adapt to unusual media to gain adequate coverage. Sometimes the medium will have to be provided by the advertisers (eg a free advertising sheet, or sponsoring a local radio station).

5.1 Press

Literacy	The first point to check when contemplating press as a media option, is the literacy levels prevailing within the country. The majority of European nations have literacy rates of 98% or 99%. In countries with low literacy, newspapers etc may only be read by the educated elite, in which case radio, TV and posters might be better solutions
National and regional newspapers	Globally there are a large numbers of national newspaper alternatives. Many countries such as the USA have more of a tradition of reading strong daily regional papers
Trade press	A strong trade press sector tends to go hand in hand with a strong economic industrial base. Less developed nations may have few, if any, trade media vehicles
'International' journals	Serious global business journals such as *Time*, *Newsweek* and *The Economist* are widely available, as are women's titles with different national editions (eg *Marie Claire*, *Cosmopolitan*, *Elle*). However, such publications are likely to be read by only a niche business and professional or lifestyle group. To reach the less wealthy, less well travelled wider populations, other local media will have to be accessed

5.2 Television

Before TV can be included on the media options list the market needs to ask: How extensive is television ownership? In the UK, 98% of households have a television. In India by comparison, television penetration to individual households is very low. However, television may be used as a group resource for several families in a village, or be shown in a public place.

There are a number of developments in television.

1   There are more satellite channels available.

2   Cable television has been introduced to many markets, offering greater choice. However, some channels cater for minority interest.

3   Digital television enables the provision of far larger numbers of channels.

A problem might be the increasing fragmentation of the audience: however, the opportunity might exist to target audiences more precisely, if discernible customer segments can be related to particular channels.

5.3 Outdoor

Posters and neon hoardings offer a good opportunity to communicate to international target groups, as copy tends to be minimal and visuals are key.

5.4 Cinema

Cinema may not always be experienced in the same way around the world. In some countries, cinema is viewed in outdoor theatres. In India adverts are rarely shown even though the audience is huge.

5.5 Radio

While the cost of purchasing a television set is prohibitive in some nations, radio ownership seems to be ubiquitous around the world. In most countries, radio plays the role of support medium and in parts of Africa such as Ghana, it is one of the most important communication mediums.

Global case study

Communicating to the Indian Consumer

The growth in media channels brought about by economic liberalization has made people increasingly aware of brand values. Advertising is of increasing concern to companies, as channels multiply and product offerings become more sophisticated. Advertising is becoming a bigger and bigger part of the marketing mix. Because markets are growing competition is increasing so you absolutely have to advertise. The biggest challenge is to identify consumer insights to make relevant advertisements. They are also concerned about the profusion of media channels – 'media clutter' – and the increasing number of brands competing for the same space. Consumer product companies also feel that a big challenge is measurement of the impact of advertising and promotional activities. The biggest issue in marketing is media fragmentation. Years ago you could run an ad on TV and know 80 per cent of TV owners would see it. Now India has so many channels it is difficult to know how to focus. There is increasing concern that advertising messages can only travel limited distances in the Indian market. A lot of companies are not able to reach further down, towards the bottom of the 'consumer pyramid'. How do you talk to someone who is illiterate, who doesn't have access to a TV or a radio? Increasingly companies will have to seek multiple advertising channels.

How To Do Business in India: GMB Publishing

5.6 The Internet

The role of the internet is discussed at length in session 10 and in the Developing Integrated Communications Strategy Study Text. Clearly the internet is a highly valuable communications tool which has changed the dynamics of the industry. It is often a cost effective medium providing opportunities for those with limited budgets.

Global case study

Customer expectations may be global, but organisations should think in terms of meeting needs at a local level. With the growth in global brands, marketers need to consider the global implications of interactive marketing using the web. Global companies quoting their customer service telephone lines on the web should not purely use US call centres with US hours of availability, or US address formats on their site. Of course marketers need also to consider using multiple languages on their site, and not just English, if they wish to appeal to a global audience. A further complication in some industries (eg financial services, telecommunications) is that regulatory and compliance issues can arise if individuals in certain countries are allowed to purchase products from a global website hosted in a specific country. Disclaimers on the home page are unlikely to be sufficient. It is another headache for global marketers and the need to manage customers to whom they cannot legally market.

5.7 Media planning and buying

As well as the usual media planning considerations (eg campaign objectives, target markets, media availability, budgets), there may be specific factors which need to be taken into account when planning in global markets. For a food product for example, media scheduling will need to take account of climatic and seasonal characteristics of individual countries.

Some international print media may be purchased from the home country market via international media representatives. Other media may have to be bought by a local country agent, acting for the advertiser. Media buying practices can vary from country to country. In the UK, published rate-card costs tend to be negotiable downwards. Other countries may be less flexible in negotiations.

Activity 2

Consider the most appropriate media that you organisation should consider in entering a new country of your choice. What are the advantages and limitations of this choice of media?

6 Other market considerations

6.1 Economic development

The level of economic development will affect the amount and way in which a product is used, affecting the message. It may also be connected with educational and literacy levels, affecting the choice of media.

6.2 Distribution infrastructure

The nature of the trading channels in the market will determine their ability to provide all or even part of the promotional effort in a market. Thus a distribution system which consists mainly of small scale market traders cannot be expected to promote the product to any significant degree, but a major distributor such as Cadbury Schweppes in the UK could be expected to promote Coca-Cola as part of the dealership. This might affect whether the objectives of the campaign are pull or push.

6.3 Competition

The level of competition will of course affect the promotional programme in a country. The entry of a foreign competitor into a market will normally provoke a response from local operators, be they indigenous or existing international companies. Thus the entrant will have to consider not only the promotional programme but his response to competition.

The above discussion suggests that promotion in global markets usually needs adaptation and benefits from expert advice.

7 Designing the global communication mix

Possible media for promotion are as follows.

- Trade and professional journals
- Consumer media (magazines, newspapers, TV, radio, posters etc)
- Direct mail
- Trade fairs and missions
- Personal selling
- Telemarketing
- Sales promotion
- Public relations

7.1 Trade and professional journals

Where they are available these publications have the advantage of providing a targeted audience. Due to the nature of the journals, literacy levels and technical expertise will be high. They provide excellent business-to-business contact but are unsuitable for consumer marketing because of their restricted circulation and technical content. Many UK professional and trade journals have significant foreign readership.

7.2 Consumer media

Here the directed nature and coverage of the media are open to question. Major media groups with global connections provide reports on readership and reader profiles. In many countries however, the data on consumer media are sparse and the global marketer has to rely heavily on local knowledge. Here are some issues to consider:

- Extent of target market coverage
- Image carried by the medium
- Literacy levels
- Ownership/readership
- Ability to convey the message
- The cost of contact (usually expressed as cost per 1,000 audience)

7.3 Direct mail

In most developed countries suitable listings based on company or consumer segmentation profiles (eg TGI or ACORN in the UK) are widely available. For trade contacts, the use of international buyers' guides provide adequate contacts. In developing or lesser developed countries the global marketer may find that such information is harder to come by. Global direct marketing, however, has been growing slowly over the last decade.

Factors driving the move to global direct marketing	<ul><li>The growth in sophistication of computer and database technology</li><li>The increasing availability of suitable consumer or business listings</li><li>The growth of international media which can be used for direct response advertising</li><li>The perceived accountability of direct marketing campaigns compared to other communications campaigns</li><li>The ease with which direct marketing campaigns can be pre-tested in order to maximise their effectiveness</li><li>The improving skills of direct marketing agencies<ul><li>– The increasing willingness of the consumer to purchase items directly</li><li>– The increasing use of internationally accepted credit cards</li></ul></li></ul>
Factors restraining the move to global direct marketing	<ul><li>Lack of telephone and postal infrastructure</li><li>Lack of road and rail penetration to facilitate distribution</li><li>Lack of suitable media to use to target consumers</li><li>Lack of consumer and business lists in some countries</li><li>The threat of increasingly strict legislation concerning the use of consumer information</li><li>Consumer backlash against what is seen as 'junk mail'</li></ul>

The major problem with using direct mail in a global context is the inability to 'follow up' enquiries generated. Thus it is important that the local sales office, agent or distributor in that country is provided with leads as quickly as possible. In developed markets direct mail can be used in several ways.

- Direct response promotion, that is buying 'off the peg'
- To generate enquiries for more personal contact
- To provide an introduction for personal contact

Generally direct mail tends to be more expensive than advertising in terms of contact costs, and has only slightly more response. However, it can have the advantage of providing a more targeted audience, which can save money significantly.

7.4 Trade fairs and missions

Trade fairs are probably one of the most effective methods of initial business-to-business contact, providing an opportunity for producers, distributors and customers to meet. They allow not only communication with a targeted audience but also the ability to demonstrate and provide trial of the product or service. Although 'fairs' exist in the consumer markets, they are less attractive and effective to the global marketer.

For smaller companies, the high cost of a presence at some of the major fairs, and the possibility of being ignored in the presence of major global competitors has led to the development of 'fringe' venues, such as a local hotel, where the company may display its offers without competitor presence. The client is often encouraged to attend by direct mail invitations and the provision of refreshments.

Advantages of trade fairs	• The ability to let potential customers see demonstrations and trials. • The ability to make personal contact with existing and potential customers to maintain and develop relationships. • Allowing direct contact with major decision makers and influencers at a time when they are interested in the product. • Providing an opportunity for market research, such as competitor activity and buyer response. • Contacting a large number of potential customers in one place.
Principal disadvantage of trade fairs	The cost, especially for smaller companies forced to compete with large multinational exhibitors. Most trade fairs are by their nature specific to a particular trade such as 'engineering' or 'toys'. Where internationally renowned trade fairs are held in your own country attendance should be considered both from the domestic and international perspective.

Global case study

The Frankfurt Book Fair is an event which brings together publishers, booksellers and customers from all over the world. Firms make new contacts and can exchange ideas.

7.5 Personal selling

In a global marketing context, personal selling will be required one or more times in the trading channel. While personal selling can be expensive is it can also be very effective. Normally advertising and direct mail can be used to generate enquiries, and the expensive resource, the salesperson, can be used to convert the lead to an order.

     Developing Global Marketing Strategy

The international salesperson cannot work in isolation but needs support.

*   Generation of enquiries
*   Product literature and samples where relevant
*   Information on price, delivery and terms

The international salesperson will require several attributes.

*   Knowledge of the product and market
*   Language and cultural knowledge specific to the country
*   Technical knowledge where necessary
*   Contacts, preferably from experience in the market
*   Suitable personality
*   Selling skills
*   Motivation to succeed

Typically there are three options available to an organisation in determining the most appropriate international sales force:

1   Expatriates: familiar with the organisation's products, policy and culture but do not have local Knowledge

2   Host country nationals: personnel based in their own country with extensive knowledge of the local market, culture and language

3   Third country nationals: employees transferred from one country to another

The advantages and disadvantages of each type of sales force are summarised below.

Category	Advantages	Disadvantages
Expatriates	Deep product knowledge	Highest costs
	Good service levels	Significant turnover
	Promotion potential following training	High training costs
	Greater home control	
Host country	Economical	Needs product training
	Good market knowledge	May be held in low esteem by stakeholders
	Language skills	
	Good cultural knowledge	Language skills need to be learnt by all
	Implement actions faster	Difficult to ensure loyalty
Third country	Cultural sensitivity	Potential identity problems
	Language skills	Blocked promotions
	Economical	Income gaps
	Enables regional sales coverage	Needs product/company training
	May allow sales to country in conflict with the home country	Loyalty assurance

Adapted from Honeycutt and Ford (1995)

Personal selling tends to be relatively more important in global than domestic markets because face-to-face contact is often so important when making overseas sales. In many Asian countries personal contact is seen as very important to selling. Sales personnel may be natives of the target market, expatriates (usually used when there is rapid market growth) or 'cosmopolitan personnel' (eg a UK telecommunications manager working for an Italian company in Russia).

The recruitment of sales personnel can present problems. This again is due to culture. In the USA, for example, sales positions are given fairly high status, while in Europe selling is seen as a low status job which will not attract high calibre people.

The idea of an 'ideal' sales person is probably outdated given the diversity of products and markets. Avon in the Asian market for example, does not cold call door-to-door, but uses personal networks of friends and relatives.

Activity 3

Explain how the role of personal selling in an overseas market may differ to its use in the domestic market.

Direct selling in this way can also be a method of overcoming the cynicism and jaded responses of consumers who are bombarded by a growing number of advertising messages and are becoming immune to sales pitches by less direct means. Research has shown that fact-to-face and word-of-mouth communication can be a more credible source of product information. On the other hand, direct selling can be seen as an unwelcome personal intrusion.

Selling industrial equipment or business services is very different to selling consumer goods. For industrial products, the use of personal selling at international trade fairs mentioned earlier may be necessary.

Personal selling can only be justified where the contribution, that is the combined effect of margin and order size, is sufficient to justify the costs involved. It is an effective tool where the market is concentrated and the product is of a high value and where demonstrating or tailoring to customer needs is required.

While it is ideal in business-to-business marketing and might be essential in dealing with government purchasing agencies, particularly in major infrastructure and industrial products it is rarely used in consumer marketing, although it can be seen when selling cars or personal financial services.

7.6 Telemarketing

Telemarketing, that is the use of the telephone to sell, is often used domestically to.

- Prospect potential customers for personal selling
- Handle repeat purchases not requiring personal visits
- Deal with customer enquiries or complaints

While widely accepted in business-to-business trading, the acceptance of telemarketing in consumer markets, especially in prospecting customers, is not so widely accepted.

Thus in the UK there is more resistance to telemarketing than in the USA. In France the existence of MINITEL has allowed the growth of telephone ordering to a considerable degree.

In the global context, the lack of a telecommunications infrastructure and cultural inhibitions to telephone selling may limit the attraction of this medium. Telemarketing requires the ability to react and thus probably works best from within a market rather than on a country to country basis.

Generally speaking, the impersonal methods of communication are less expensive in terms of cost per contact, but less effective than personal communication. The order size and potential contribution may well dictate the appropriate medium. Thus high value, high volume sales on a business-to-business basis warrant trade fair visits, and foreign sales representation, whereas consumer communications are better suited to indirect methods such as advertisements.

7.7 Sales promotion

Sales promotion describes a range of techniques appropriate for targeting consumers, for instance via price reductions, free gifts, or competitions. Trade and sales force promotions are also implied under the general heading of sales promotion.

Different countries have their own local restrictions concerning different sales promotional devices. For instance, collector devices are a well used method of encouraging repeat purchase in the UK. However, they may not be allowed in others such as Germany, Luxembourg, Austria, Norway, Sweden and Switzerland.

Sales promotions that are to run across a number of different countries can only succeed by tapping into common tastes, interests and activities.

7.8 Public relations

Public relations is 'the planned and sustained effort to establish and maintain goodwill and mutual understanding between an organisation and its publics' which include:

- Customers
- The local community
- Shareholders
- Media
- Employees
- Government
- Suppliers
- Pressure groups
- Trade intermediaries

There is wide disparity in the sophistication of PR from country to country, as well as the usual list of cultural, language, media and legal barriers. The political context of individual countries may also affect the extent to which different public relations techniques can be used.

7.9 Sponsorship and celebrity endorsement

Sponsorship as a method of international marketing is burgeoning. Nike sponsored the Brazilian football team in the World Cup, and backed this up with poster and television campaigns. Nationwide, a British financial institution, sponsored the England football team until July 2010.

Every major football club in the UK (and certainly those which have a high international profile) are sponsored by major companies. Formula 1 motor racing, a multi-million pound industry, is heavily sponsored by international tobacco companies, although this is to be phased out.

Organisations are keen that they sponsor events or organisations with a good public image. Johnson & Johnson withdrew from sponsoring the 2002 Winter Olympics in Salt Lake City, partly because of the corruption and bribery scandal surrounding the International Olympic Committee.

Global case study

Anheuser-Busch InBev unveils Budweiser World Cup marketing plan

There's always been a Manchester United, a West Ham United and a Leeds United, but with Budweiser United, Anheuser-Busch InBev is looking to add a new name to those storied clubs and promote its FIFA World Cup sponsorship.

All of the brand's global marketing for this summer's event will use the "Budweiser United" sponsorship logo, which is a red, soccer-style shield with the World Cup trophy. The logo is the centrepiece of a multi-dimensional campaign that includes interactive, digital and grassroots components.

A-B InBev Chief Marketing Officer Chris Burggraeve said the campaign is *"about uniting beer drinkers around the world and celebrating the game of football."*

The World Cup activation is the most significant global marketing plan A-B InBev has undertaken with the Budweiser brand since InBev acquired St. Louis-based A-B for $52 billion in 2008. It also shows how A-B InBev will use Budweiser's largest global sports sponsorship to push more brands within the InBev portfolio. For the first time during a World Cup, the company will leverage its sponsorship to push Brahma in Brazil; Hasseröder in Germany; Jupiler in Belgium and The Netherlands; and Harbin in China. Burggraeve called the change *"a win, win, win,"* and added, *"It's a win for FIFA because their rights are connecting in a relevant way, it's a win for brands that can now really bring the property to fans, it's a win for us because we're getting a higher return on investment."*

In an effort to connect more directly with fans, A-B InBev will change its "Budweiser Man of the Match" program. Winners of the award will be selected by fans who vote on FIFA.com. The company also worked with FIFA to tweak the program so that a fan chosen through a Budweiser sweepstakes – rather than a FIFA official – will present the award to players at the end of games. Budweiser has enlisted artists in South Africa to design trophies to be presented to the "Man of the Match" after each of the 64 World Cup games this summer.

"We want fans to be able to connect with the game like never before," Budweiser Global Director of Advertising Andrew Sneyd said.

Budweiser also will create a "Real World"-style, digital reality show called "Bud House." The company has secured a house in South Africa and plans to put 32 soccer fanatics (16 men and 16 women) representing each participating World Cup country under one roof. The fans will bunk based on the group stage, so that the U.S. fan will share a room with a fan from England. Six to eight segments of the show will be posted online each day, and the winning fan of the winning team will be able to award the "Budweiser Man of the Match" trophy to the top player in the World Cup final. Budweiser Global VP Jason Warner said, *"This will be our biggest program."*

http://stlouis.bizjournals.com/stlouis/stories/2010/03/22/daily58.html

Accessed 27 March 2010

8 Planning the global marketing communications campaign

The way in which an organisation handles a global marketing communications campaign will depend upon a variety of factors:

- The number of markets in which the company operates and the relative importance of sales in those markets

- The extent to which the different markets are similar or dissimilar

- The way in which the company is currently organised to do business internationally

- The corporate culture and management style of the organisation

- The degree to which the company has a corporate world-wide identity

- The skills and money available in the different markets

In planning a global marketing communication campaign the global marketer needs to make decisions on the following:

- The marketing communications strategy ('push', 'pull' or 'profile')
- Professional assistance
- The message
- The media
- The promotional budget
- Monitoring and control
- Organisation
- Independent or co-operative promotion

8.1 Professional assistance: selecting an agency

The normal form of assistance is to hire an advertising or marketing communications agency. This will be either of the following.

- Global agency with local offices
- Local agencies in each market

Considerations when making the decision include:

- The agency's knowledge and coverage of the relevant market
- The agency's quality and reputation within each market
- The additional services provided (for example, market research, public relations etc)
- The ability to liaise with the agency easily
- Whether the campaign is to be standardised
- The value of the promotional budget
- The organisation of the global marketer's organisation

8.2 Selecting the message

While commonality of needs is largely recognised worldwide, the way in which these needs are satisfied varies, as does the frequency with which these needs occur in any society.

Social and cultural values will almost inevitably mean that selecting the message will require local advice, either from the local office or the local distributor. Further considerations concern the following.

- Localised or standardised campaigns
- Market conditions
- Market segments sought

The degree to which promotion should be standardised is a difficult decision. Where the product or service is perceived or used in a significantly different way, the message has to be adapted. Where the media availability and market coverage is significantly different from the home market, a new and different media campaign will be required. Finally, legal restrictions may force both message and media adaptation.

Advertising guidelines dictated centrally may be easy to implement and control, but local agencies can rebel against tightly delineated operating rules. Local staff may feel that a standardised execution is not correct for their particular market.

- The 'not invented here syndrome' can be overcome by involving regional subsidiaries at the initial strategic stages of a piece of work, so that everyone has a sense of ownership over the finished campaign

- If advertising is totally adapted to meet the individual circumstances of the local marketplace, regional subsidiaries are likely to have high involvement in the campaign

- Pattern advertising is a halfway house between total standardisation and total adaptation.

  - Here, agency and client head office dictate the strategic direction which the advertising must take. Local subsidiaries interpret this to suit the specific characteristics of their market.

  - A slight variation to pattern advertising is offered where head office dictates the strategic direction and produces the bulk of the creative work centrally. However, the advert is finished off with local shots.

Customer expectations may be global, but organisations should think in terms of meeting needs at a local level. With the growth in global brands, marketers need to consider the global implications. An American Express pan European television campaign featured two different executions. One was a picturesque country wedding; the other a Paris fashion show. The key wedding and catwalk sequences were common to all countries in which the adverts ran. However, scenes involving card members were shot locally in order that each market could feature authentic backgrounds and nationals playing the main roles.

8.3 Selecting media

As suggested in the discussion above, the availability, cost and effectiveness of media may not be the same as in domestic markets. In considering the use of any media one needs to take into account the following:

- The target market you wish to contact
- Degree of market coverage
- Legal restraints on using the medium
- Physical constraints of medium (eg ability to show colour etc)
- Degree of customer response to the medium
- Cost per enquiry generated

Generally, unless your organisation is a multinational company operating in a particular country, such information will be difficult to obtain. The advice of either your local distributor or advertising agency should therefore be sought.

8.4 Deciding on the promotional budget

As in domestic markets there are several methods used to decide on international promotional budgets. The first two are the most common. Unfortunately they do not take into account either competition or the requirements to achieve sales objectives.

- What can be afforded
- Percentage of sales
- Competitor parity
- Historical precedent
- Objective funding

Objective funding assumes the following.

- Sales objectives in a particular market can be quantified
- The appropriate level of promotion can be determined
- The necessary funding will be provided

This is a more appropriate method of deciding on promotional budgets, and has the advantage that when sales are hard to obtain, a case can be made for an appropriate budget. In practice, promotion is often regarded as an 'optional' cost and considerable diplomacy is required by the global marketer to justify increased budget at a time of financial stringency.

8.5 Monitoring and control

Investment of any kind should be related to the return that the investment generates, and this applies equally to promotional campaigns. Costs are usually easy to obtain from the internal records. Typical monitoring measures could include the following.

- Coding advertisements so that enquiries can be related to a particular medium or advertisement

- Comparing response rates between media

- Comparing against other salesforces, calls and orders per call to establish the costs of contact and order generation

As a result of this and similar feedback the mix can be adjusted to improve promotional productivity. Modern information technology using database and management information software has reduced the effort required to monitor programmes.

8.6 Organising international promotions

In most cases the organisation will rely on the advice and help of both the local distributor and agency for operational and administrative decisions. Strategic decisions will always be the domain of the company.

Part of the task of a channel of distribution is that of promoting the product. Trade promotion is usually undertaken by the producer in most markets with enquiries being direct to the distributor or agent. Where consumer markets are involved the main distributor is usually required to provide the promotion.

Smaller distributors such as retailers may well be tempted to promote the product locally to the end users, if an incentive is offered. Thus a foreign department store may be tempted to put a special promotion on for a product if the costs can be shared between it, the distributor and the producer.

9 Global agency selection and management

9.1 Global agencies

As organisations have expanded their operations globally, so too have advertising agencies. Many of the large agencies have developed globally, either by setting up branch offices in foreign countries or by merging with or acquiring local agencies.

Some agencies expanding abroad prefer to establish global networks or alliances where local offices are not wholly controlled. The argument is that local partners with a stake in the agency will be motivated to produce superior work.

Media independents have mirrored the pattern of agency development and many belong to international media planning and buying groups. The trend among clients is towards the centralisation of advertising. Many large companies believe international brands are best served by an agency operating internationally.

9.2 Selection

Selecting an agency will follow a series of well defined stages. Locating suitable agency candidates is the first step in the process:

- Initial search. Prospective clients will probably be aware of the large multinational agencies based within their own country.

- A shortlist of agencies will then be drawn up, usually on the basis of their current work and past track record.

- The client will then visit the local offices of those agencies for a series of credentials presentations. These initial visits will help to form an opinion about which candidates should be requested to formally pitch for the client's business. All agencies involved in the pitch should be given the same written brief to follow.

- The agency's response to the brief will usually involve a formal presentation backed by a written proposal document with several important features.

    - The agency's interpretation of the client's advertising problem
    - The creative and media strategy which will ensure objectives are met
    - Control mechanisms to be used
    - Timing schedules
    - Allocation of responsibilities
    - Costings
    - Terms and conditions of business

- The final selection decision will have to take into account many client side factors such as the client's organisational structure and management style, the number of brands to be advertised and the degree to which brands penetrate different markets.

- Other criteria for selection would include the following.

    - The types of advertising and other communications services offered

    - Level of expertise in the client's field of work

    - The agency's global creative track record

    - The balance within the agency of campaigns handled for local clients and those handled for global ones

    - Whether the agency has strong local offices in the client's home and other key markets

    - The extent to which the agency's culture and management style fits with that of the prospective client

    - The potential conflict with existing business handled within the agency network

    - The control and co-ordination procedures in place

Activity 4

Your organisation is about to launch a new product in the Middle East. What sorts of assurances would you want from an advertising agency pitching for this account?

9.3 Advantages and criticisms of using a global agency

The advantages of using an international agency

- Less duplication and dilution of effort on the part of agency and client
- Centralised control of all advertising effort
- Speedy response across markets
- Pooling of talent and ideas from the entire agency network
- Specialised resources available
- Standardised working methods by the agency
- Reduced costs due to economies of scale

- They provide an uneven quality of service in their different branches
- They produce bland campaigns
- Quality control suffers due to handling hundreds of campaigns simultaneously
- Clients have to tailor their campaigns to suit the conventions of the agency
- Small or medium sized clients suffer a lack of attention from senior staff
- High staff turnover rates exist among creative employees

Despite the increasing presence and power of global advertising and media networks, independent local agencies exist in the markets of most countries. Some companies prefer to retain country by country agency arrangements, believing local agencies to be creatively closer to their own markets.

Picking local agencies is an appropriate strategy for the client who has only a small portfolio of products to be advertised in a limited number of markets.

Global case study

In 1999, advertising industry leaders formed what is now the Marketing and Advertising Global Network. Today, MAGNET is a group of non-competing, independently owned advertising agencies in major markets throughout the world. The network is comprised of leading agencies, with billings ranging from 8 million to 200 million US dollars, sharing a desire to pursue and achieve excellence in this profession of marketing. MAGNET provides a way for member agencies to share their experience, knowledge and ideas with other agencies in other parts of the world. Working together, MAGNET agencies provide an easily accessible network of information and financial skills used to enhance agency operations and improve the bottom line.

http://www.magnetglobal.org

9.4 Management issues

External factors, such as market diversity, segmentation and competition affect the type of co-ordination chosen. Internal factors, unrelated to the market, can also dictate the management and co-ordination of global marketing communications activity.

1   **Organisation structure**

   (a)   **Local autonomy.** Each subsidiary of an international agency may act as a separate profit centre, attracting its own clients in the home market. Upon appointment, the subsidiary which has brought in the client takes the role of lead agency office, with overall supervision of the client's account

   (b)   **Central control.** Alternatively, an agency may exert strong central control on regional offices from its headquarters base.

2   **Organisation culture.** Managers may have different assumptions as to how advertising ought to be done, and it might be a basic hidden assumption that decisions are taken at the top or, on the other hand, by giving local managers their lead.

3   **The need for integrated marketing communications.** Organisations with worldwide exposure may need central control of marketing communications to ensure they are, in fact, integrated.

9.5 Control

9.5.1 The advertiser

A number of management and control problems need attention:

1. Budgets. If advertising budgets will be split among the different units, the money might be spent less effectively than if control is centralised.

2. Timing. The local advertising campaigns might need to be co-ordinated so that the production side can cope with demand. This is especially true if the firm is an exporter, with a number of local sales offices and warehouses. Local advertisers may generate demands which cannot be met.

3. Some central review is necessary so that good ideas can be passed around within the group.

4. Expertise. The person responsible for advertising in a 'small' subsidiary may not have a great deal of expertise and may rely too much on local advertising agencies for ideas, rather than controlling the output directly.

5. All advertising campaigns need to be evaluated for effectiveness.

6. Local campaigns may have different objectives, and so appropriate effectiveness measures need to be outlined.

7. It is easier to measure message recall and media buyer efficiency in advanced economies, and there may not be the facilities in less developed countries, even if these promise high growth.

9.5.2 Between the client and the agency

Relationships with agencies need to be managed. In many agencies, there will be an 'account manager' responsible for overall liaisons with client. It is in the agency's interests to establish a long-term relationship with the client.

The agency's account executive is critical in communicating the client's needs to the creative team. The agency must also be able to liaise, where necessary, lower down the chain of command, and to communicate to the units of the business, in clear terms, the objectives of the promotion campaign.

9.5.3 Within the agency

Within the agency, there are the following issues of management and control:

1. A general problem of central direction versus local freedom. This we have already discussed, in terms of tailoring local advertising campaigns or adopting a standard approach.

2. Conflicts of interest: how free are local offices to tout for business, if this involves offending a major client?

3. How do you co-ordinate the activities of different creative teams?

4. There may exist problems of corporate culture, particularly if an agency grows up as the result of a takeover.

5. Performance appraisal and culture.

6. Many human resources issues are relevant (eg corporate culture, details of organisation structure).

9.6 Current client and agency issues

As markets expand, clients are likely to forsake traditional, vertical organisational structures where brands are managed on a country by country basis, in favour of a horizontal structure which cuts across country divides.

This implies brand and product management at a centralised level and may result in a preference for centralised agencies. Consequences might be as follows.

1 Clients are likely to become more demanding of their agencies, as clients strive to ensure that their advertising is accountable and effective. In America, the trend is already towards payment by results. Clients are also likely to demand a larger base of expertise in terms of communications and research services provided.

2 Agencies will continue to expand globally to meet the needs of their clients. This may lead to a concentration in advertising agency ownership as the large agencies seek to expand still further by way of acquisition and merger.

3 Agency expansion will also mean that an increasing number of local agency offices will be established in new markets (eg Russia, Eastern Europe, China). Agencies may need to take a long term perspective on emerging markets.

Activity 5

Critically evaluate the case for using one advertising or marketing communications agency to create and implement an international advertising campaign for your organisation.

1 Consider the issues in global marketing communications

- The global marketer needs to decide whether or not it is appropriate to standardise communications across markets.

2 Understand the cultural considerations in global marketing communications

- Organisations operating outside their home markets need to be aware of the implications of cultural differences for all aspects of the marketing mix.

- Verbal and non verbal communications, aesthetics, dress and appearance, family roles and relationships, beliefs, learning and work habits are dimensions of culture of particular relevance to communications.

- Planning and media buying across borders can be a complex task. Media availability can vary greatly from country to country. Media conventions which apply in a home market may not apply elsewhere.

3 Consider the legal aspects in global marketing communications

- Laws and regulations governing marketing communications must obviously be observed. Each country will have its own set of restrictions which apply to advertising, packaging, sales promotion and direct marketing.

4 Consider the media aspects in global marketing communications

- Clients and their agencies can choose to handle global advertising campaigns in a number of ways. Although a variety of factors will influence the management of any particular campaign, the organisational structure of the client company will play an important role.

- Some agencies have expanded abroad by setting up their own subsidiaries overseas. Others have established alliances with local agencies already in existence.

- There are arguments both for and against using the services of a global advertising agency. The current preference among large clients is to centralise advertising with an internationally based agency, rather than choose local agencies on a country by country basis.

5 Understand the wider market considerations

- The standardisation versus adaptation debate has implications for agency management. If campaigns are totally standardised across markets, the lead agency will take the major role in designing and implementing the global campaign; local agency subsidiaries will have minimal input.

6 Evaluate how to select global agencies and manage them successfully

- The process of selecting a global advertising agency involves initial search; credentials presentation and shortlist; competitive pitch and final selection. The client will gauge competing agencies against criteria such as response to the brief; types of communication service provided; expertise in handling local and international campaigns; and similarity of management style and culture to that of the client.

1   The cultural issues will depend on your organisation, the product and service you are offering and the country you are entering. It will be essential to be well informed and base your appreciation of cultural issues on local knowledge of the marketplace.

2   The media selection will depend on your organisation and the country of selection. Countries differ in literacy rates and media preference. For example in Ghana many people listen to the radio and have the television on at work and these media are considered more important than print media.

3   Personal selling differs by market. The role of personal selling will differ according to the organisation's objectives, financial resources and the attitudes towards personal selling in a given market.

4   When selecting an agency for a specific market you will need to be sure that the agency meets closest the following criteria (a non-exhaustive list):

- Level of expertise in the product sector and country
- The agency's track record in the product sector and country
- Understanding of the cultural, political and legal issues
- Potential conflict of interest with other competing brands
- Existing relationships with key media

5   In critically evaluating the case for using one advertising or marketing communications agency to create and implement an international advertising campaign for your organisation you will need to consider the following issues (a non-exhaustive list):

- Degree of control required at Head Office in controlling the brand and the message
- Degree of standardisation required
- Availability of local agencies with the expertise required
- Availability of resource to manage and co-ordinate the campaign

Hollensen, S., (2007). *Global Marketing*. 4th edition. London: Prentice Hall.

Honeycutt, C.D., and Ford, J.D., (1995) '*Guidelines for managing an international sales force*'. Industry *Marketing Management* vol 24, p.138

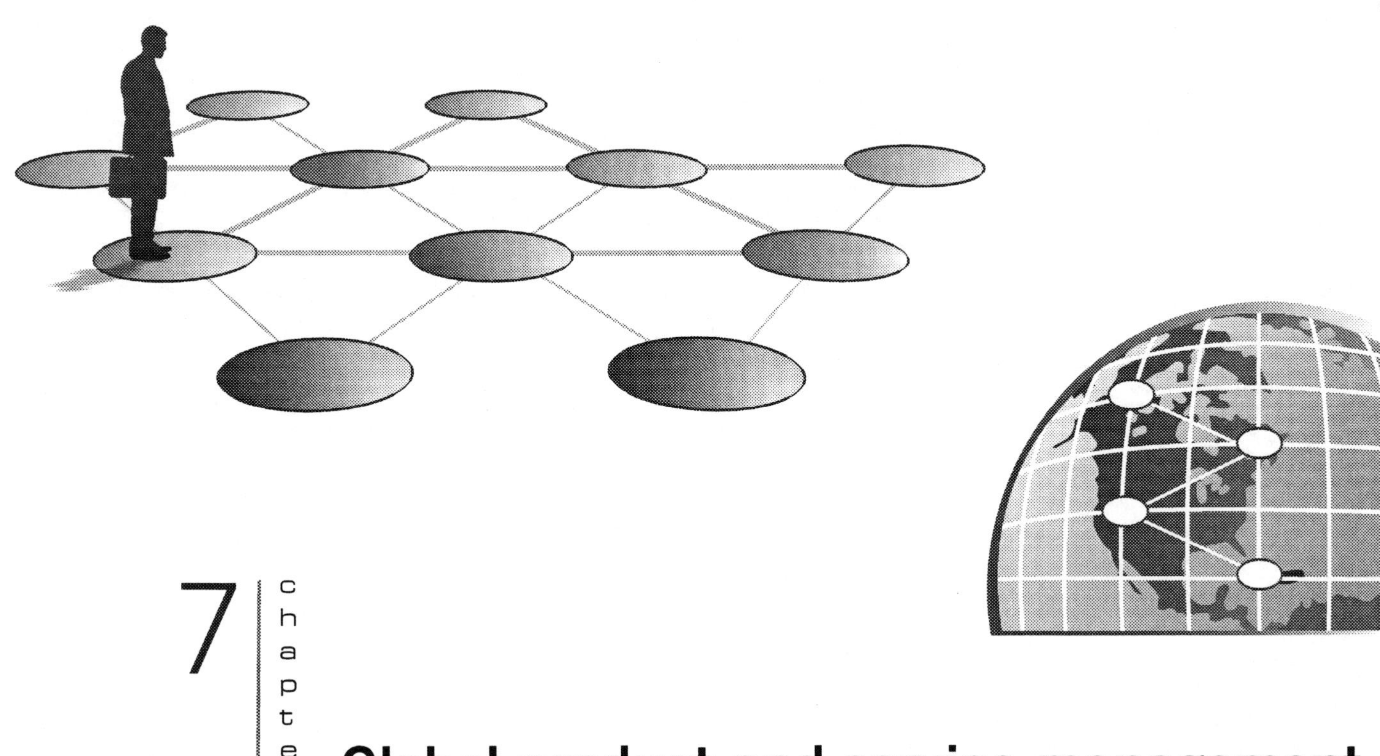

7 | chapter

Global product and service management

The success of the global marketer depends ultimately on developing products or services that meet the demands of the market, profitably. With shortening life cycles and fewer differentiated products the global marketer is under more pressure than ever to meet the needs of the global consumer.

The purpose of this session is to focus on recent trends in global product and services marketing and identify how products and services can be managed and developed successfully in a global context.

We begin by looking at what a product and a service actually is and understanding the various benefits that make up the product offer. We then consider the factors that encourage either product standardisation or product adaptation in the context of global marketing. We consider the role of the product life cycle in managing products across borders, the role of the international product life cycle concept and product portfolio analysis in the management of global products. Finally we explore the new product development process for global markets and decisions that need to be taken in deciding on *global* branding strategies.

Contents

Chapter learning outcomes

In this chapter you will cover the following:

- Explore global product and services marketing
- Consider the balance between standardisation and adaptation
- Evaluate global product policy
- Assess the management of products across borders
- Evaluate new product development for global markets
- Consider global branding decisions

1 Setting the scene

Product and service decisions are among the most critical decisions that a global marketer must make and sets global marketing apart from domestic marketing probably more than any other aspect of the marketing mix.

The focus of this chapter is to explore product and service-related issues and explore approaches for handling them. It explores the issue surrounding the choice of whether to standardise or differentiate the product offering by adapting it to the various markets in which the organisation seeks to operate. On the one hand, if you are expecting to achieve economies of scale, the last thing you want to do is customise each product. On the other hand, many products have to be adapted for overseas markets.

This adaptation does not only relate to legal and technical requirements, but also to the different cultures in each market. This applies also to augmentations of the product, such as packaging, labelling and servicing. The development and introduction of new products also is influenced by global marketing considerations, in particular the fact that product life cycles can differ in each country. Particularly important is the need to provide customers around the world with a satisfactory experience.

Assessment advice

Continue to identify the major environmental trends in global trade. Consider the ways in which these trends have affected the product and service development and management across the global markets they operate in and the influences it will have on the markets it is seeking to operate in. Gain a clear understanding of the way the organisation you are investigating views the world and identify examples that give you clues to its global product and services marketing strategy. In an ideal world the organisation might like to manufacture a standardised product or deliver a standardised service. Consider the case for standardisation versus adaption and the circumstances which support or prevent its implementation.

2 Global product and services marketing

2.1 What is a product?

Definition

A **product** is anything that can be offered to a market for attention, acquisition, use or consumption that might satisfy a want or need. It includes physical objects, services, places, organisations, ideas, and people.

Developing Global Marketing Strategy

In creating an acceptable product offer for the global market it is necessary to firstly examine what constitutes the 'total' product offer. As a reminder, a product exists at three levels - core, formal and augmented.

2.1.1 Core product

The core product refers to the need which is being satisfied or the problem which is being solved by the product. This is a vital concept in global marketing, (eg the 'core' product might vary between developing and developed countries).

2.1.2 Formal product

The formal product is what the market recognises as the tangible offer. It comprises the features, styling, quality, packaging and brand name. Again this may have important implications for international marketing. Variations in quality, size, colour and so on may have to be used from country to country.

2.1.3 Augmented product

The augmented product includes core and formal products and the additional services and benefits which 'surround' a product. They may offer the following.

• Reputation	• Guarantee
• Delivery	• Finance
• Before and after-sales services	• Credit
• Installation and maintenance	

So it is vital for the international marketer to be aware of needs and expectations and the extent to which they vary between different countries when making product decisions.

2.2 The dimension of the global product offer

Similar to the above model, in creating an acceptable product offer for global markets Kotler (2002) suggests three essential aspects of the product offer which should be considered.

1   Product benefits
2   Product attributes
3   Marketing support service

This is shown in the following model.

The three levels of a product

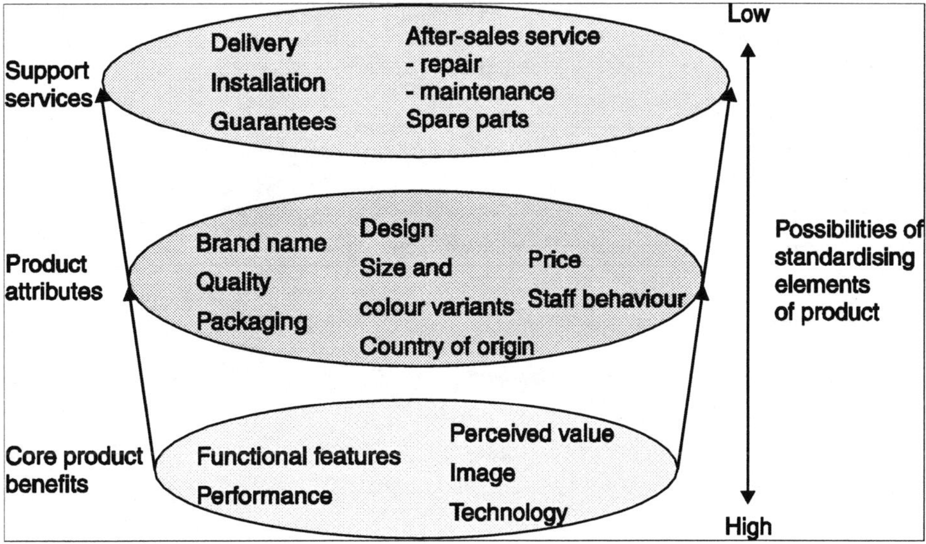

We can see from this model that it is much easier to standardise the core product benefits than the support services, which often have to be tailored to the business culture and sometimes to individual customers.

This is a question that the global marketer has to consider seriously. If only language has to be adapted then the impact on overall cost will be minimal but if more fundamental changes to the product are required, because of differences in safety regulations or standards, then the higher cost might prove prohibitive.

Such problems can be circumvented by adapting the market entry approach for any given market but in making such strategic decisions the organisation has to be clear exactly what product or aspect of the augmented product will be most valued by customers in each global market.

2.3 Services

Services have the following characteristics.

- Intangibility: elements that cannot be touched, smelled or seen
- Inseparability: many services are created at point of sale
- Variability/heterogeneity: dependent on who delivers service, and when it takes place
- Perishability: services cannot be stored

There are a number of specific problems in marketing services internationally:

- Achieving **uniformity** of the different marketing parameters in remote locations where exercising control can be problematic
- **Pricing**, as fixed costs can be a significant part of the total service costs
- Consumers' **ability to pay**
- Perceptions of **service**
- Preserving customer **loyalty**

The marketing mix for services also includes people, processes and physical evidence but there are issues of overall service decision, quality and delivery and management.

(a)   **Overall service design**. Store design and layout may need to be adapted to meet local tastes and preferences.

(b)   **Service quality and delivery**. Service improvements in one country can be introduced to another country, providing the market is open.

(c)   **Service management**. Service companies in one country can expand by buying up service facilities elsewhere, to provide resources of finance and expertise.

These have a number of implications for the marketing mix.

People are a key element in service delivery. Firms providing international services with high customer-contact, such as airlines, have been well aware for many years of the need to motivate service staff to give their best. For example airline cabin crew can be trained to speak the languages relevant to the routes they fly.

Processes describe the way the service is delivered. Of course, this is infinitely bound up with people, but well designed processes are essential to service design and delivery. Expectation of standards may vary with different cultures and therefore standardisation is difficult to achieve. Physical evidence in international marketing terms can include factors such as these.

(a)   Coherent branding across service outlets in the world. This will be reflected in packaging (of materials supplied with the service), staff uniforms and so on.

(b)   Attention to the design, layout and lighting of any physical environment, such as aircraft cabins, hotel rooms, restaurant.

(c)   People communicate both by using language and non verbal signs which differ around the world.

Because of the significance of interpersonal relationships in services marketing it is often cultural empathy in the way services are developed and delivered that is critical to commercial success.

Global services marketing is a major growth area. Using one service sector as an example, consider the main barriers to success and what strategies might be deployed to overcome them.

3 Standardisation and adaptation

There are a number of factors that affect the global management of products and services including the balance between standardisation and adaptation.

3.1 Factors encouraging product standardisation

Factors encouraging standardisation are as follows.

(a) Economies of scale

- Production
- Marketing/communications
- Research and development
- Stock holding

(b) Easier management and control.

(c) Homogeneity of markets, in other words global markets available without adaptation (eg Nike trainers).

(d) Cultural insensitivity towards certain products (eg lingerie)

(e) Consumer mobility that means standardisation is expected in certain products.

(f) Where 'made in' image is important to a product's perceived value (eg France for perfume, Italy for coffee machines).

(g) For a firm selling a small proportion of its output abroad, the incremental adaptation costs may exceed the incremental sales value.

(h) Products that are positioned at the high end of the spectrum in terms of price, prestige and scarcity are more likely to have a standardised mix.

The decision for most organisations to standardise or adapt is based on a cost-benefit analysis of what they believe the implications might be for revenue, profitability and market share. Only if the needs and tastes identified are significantly different and substantial business will be generated is there any justification for delivering an adapted product.

Some believe that continual exposure to standardised products will redefine customer needs and ultimately lead to a change in tastes, thus leading to greater market share in the long term.

Generally the organisation benefits from adopting a standardised strategy through:

- More rapid recovery of investment
- Easier organisation and control
- Reduction of costs through economies of scale
- Experience improving the organisation's operations

Some disadvantages of standardisation include:

- Opportunities may be lost when it is not possible to match specific local market requirements
- Mangers of local subsidiaries may become de-motivated if they are not able to innovate locally
- Competitors find it easier to copy standardised offerings

3.2 Factors encouraging product adaptation and modification

3.2.1 Mandatory product modification

Mandatory product modification usually involves either adaptation to comply with government requirements, legal standards or unavoidable technical changes.

3.2.2 Discretionary modification

(a)   Discretionary modification is called for only to make the product more appealing in different markets. It results from differing customer needs, preferences and tastes. These differences become apparent from market research and analysis, intermediary and customer feedback and so on.

(b)   Levels of customer purchasing power. Low incomes may make a cheap version of the product more attractive in some less developed economies.

(c)   Levels of education and technical sophistication. Ease of use may be a crucial factor in decision making.

(d)   Standards of maintenance/repair facilities. Simpler, more robust versions may be needed.

(e)   'Culture-bound' products such as clothing, food and home decoration are more likely to have an adapted marketing mix.

Global case study

Specific problems have been faced by fast-food restaurants as they have roll-out their global plans. In India issues such as kosher and halal meat supplies have to be considered. Also ingredients such as beef and pork will prove unacceptable to many consumers in India. The necessary ingredients for fast food, such as wheat for pizza bases, suitable chicken and so on are unavailable in certain countries. Indian consumers prefer a variety of foods and so Pizza Hut and KFC are often located under one roof.

Ideally, an organisation prefers to offer the same product, with the same pricing policy using the same promotional methods and through the same distribution channels in all its markets, for ease of management, if nothing else. But in practice, this is seldom ever possible.

At the other end of the spectrum, it has been argued that adaptability is the key ingredient for global success. Much of the decision making as a global marketer is concerned with taking a view on the necessity, or lack of it, of adapting the product, price and communications to individual markets.

Discussion

Study buddy

Share your thoughts with your Study buddy as to the advantages and disadvantages of standardisation and adaptation for the organisations you are investigating for your assignment. Together then share with the wider cohort through the discussion forum the conclusions you have reached.

An organisation's approach to this decision depends to a large extent on its attitude towards globalisation and its level of involvement in global marketing. There are broadly three types of approach in this context, as discussed earlier:

Ethnocentrism	Overseas operations are viewed as being secondary to domestic operations and are often simply a means of disposing of surpluses. Any plans for overseas markets are developed at home with very little systematic market research overseas. There is little or no modification of any aspects of the mix with no real attention to customer needs. This is the first step into international marketing and involves a centralised strategy.
Polycentrism	Subsidiaries are established, each operating independently with its own plans, objectives and marketing policies on a country by country basis. Adaptation will be at its most extreme with this approach. Polycentrism can be viewed as an evolutionary step and involves a decentralised strategy. It is easy to fall into a multi-domestic pattern of operations that does not take advantage of co-ordinating actions across differential markets.
Geocentrism	The organisation views the entire world as a market with standardisation where possible and adaptation where necessary. It is the final evolutionary stage for the multinational organisation and involves an integrated marketing strategy.

In general terms, the extent to which the mix has to be adapted depends on the type of product. Some products are extremely sensitive to the environmental differences, which bring about the need for adaptation; others are not at all sensitive to these differences, in which case standardisation is possible.

A useful way of analysing products globally is to place them on a continuum of environmental sensitivity. The greater the environmental sensitivity of a product, the greater the necessity for the company to adapt the marketing mix.

A more sophisticated approach is a two dimensional matrix. The vertical dimension measures the advantages of standardised marketing and the horizontal dimension takes the need for localised marketing into account.

<table>
<tr><td colspan="2">

Sector 1
GLOBAL

- Aircraft manufacturing
- Computers
- Industrial machinery

Automobiles

</td><td colspan="2">

Sector 3
BLOCKED GLOBAL

- Telecommunications
- Generators

Pharmaceuticals

</td><td>High</td></tr>
<tr><td colspan="2">

Sector 2
MULTINATIONAL/ MULTIMARKET
- Medical equipment
- Synthetic fibres
- Cash dispensers
- Electrical equipment

</td><td colspan="2">

Sector 4
NATIONAL/ LOCAL

- Breweries
- Cement
- Retail trade

Processed food

</td><td>

Advantages of standardised marketing

Low

</td></tr>
<tr><td>Low</td><td></td><td>High</td><td></td><td></td></tr>
</table>

Need for localised marketing

(a) **Sector 1** contains true global marketing companies with a geocentric orientation. Local adaptation is inappropriate and globalising forces can be exploited to great advantage to the company. Examples include aircraft, computers and industrial machinery.

(b) **Sector 2** Multinational or multimarket companies with a polycentric orientation adopt this approach. Products require only a low degree of local adaptation. The world market for such multinational organisations is divided into regions or countries with different characteristics, such as W Europe, S America or the Far East. Products in this sector include electrical equipment.

(c) **Sector 3** Blocked global businesses are those in which both the need for local adaptation and the globalising factors discussed above are strong. This sector includes businesses that are dominated by economies of scale and would be global but for the influence of legal or political constraints (eg government purchasing policies) creating the need to adapt their products. Regional telephone networks offer a typical example.

(d) **Sector 4** contains true local businesses. Strong local adaptation is necessary for success and there are no strong arguments in favour of globalisation (eg brewing and retail trade).

Global case study

Most radios are powered by batteries, or by mains electricity supply. Batteries are expensive, and many rural areas in poorer countries do not have mains electricity. Whereas in developed nations broadcasting is a source of entertainment, in lesser developed countries it can be a vital source of education and modernisation. Invention of a clockwork radio has been credited with having brought broadcasting within reach of many people. The clockwork ratio converts kinetic energy into electricity - the spring powers a tiny generator, which powers the radio. It is hard to think of a product better adapted to poorer markets. An example of the radio's use is in health education in Eritrea (reconstructing itself after a war of independence from Ethiopia which has lasted several decades). Villagers can learn about simple ways of reducing children's exposure to infection.

4 Global product policy

4.1 Global product portfolio

The decision about which products should be included in the range to be marketed globally is determined by a number of factors:

- Organisational objectives in terms of growth and profits

- Philosophy towards global development

- Which financial and managerial resources will be allocated to global marketing

- Characteristics of the markets, such as economic development and barriers to trade, for example

- Requirements, expectations and attitudes of the consumers in the market

- The products and services, their attributes, perceived values, their stage in the life cycle, economies of scale and so on

- Support products required from other elements of the marketing mix

- Political, legal and other environmental issues that must be overcome

- Level of risk that the organisation is prepared to take.

4.2 Global product strategies

Organisations adopt a wide variety of product strategies in global markets. Mesdag (1985) suggested that an organisation has three basic strategic choices:

SWYG **Sell What You Have Got**	• Most common form of export strategy
	• Focus on a few global markets as the main objective is to fill production in the domestic market
	• Most successful global products started off as domestic products with such a strategy and particularly where such products can be positioned as trans-national/cross-border brands based on identifying and meeting the needs of customer segments
SWAB **Sell What People Actually Buy**	• Only possible to penetrate one market at a time
	• Difficult to compete with local organisations on their own terms
	• Difficult to establish credibility where there is strong domestic demand and brands
	• Makes considerable demands on the organisation's resources and therefore is often impractical
GLOB **Sell the same thing GLOBALLY** **disregarding national frontiers**	

Keegan (1995) highlights the key aspects of global marketing strategy as a combination of standardisation or adaptation of product and promotional elements of the marketing mix. He offers five alternative approaches to product policy.

Strategy	Comment
Straight extension	• Introducing a standard product with the same promotional strategy
	• Achieves major savings in market research and product development
Promotion adaptation	• Leaving a product unchanged but adapting the promotional activity to take into account cultural and other differences.
	• Relatively inexpensive and cost-effective strategy
Product adaptation	• By modifying only the product the organisation maintains the core product in different markets but maintains the promotional message (eg McDonald's 'I'm loving it' campaign remains unchanged around the world)
Dual adaptation	• Adapting both the product and promotion for each market to take a totally differentiated approach.
	• Often adopted when the organisation is a market follower or new to the market and can be an more expensive but often necessary strategy
Product invention	• Adopted by organisations from developed nations that supply product to less developed nations
	• Existing products may be too technologically sophisticated to operate in less developed nations where power supplies may be intermittent and local skills and capabilities may be limited

GMN expert view

Neil Stevenson – module advisor

"The whole spectrum from extension to invention is crucial in my view. A wrong decision here creates a domino effect across the entire marketing strategy/mix and can lead to high costs being incurred on unsuccessful product launches."

4.3 Other aspects of product decisions

Other aspects of the product decision principally concern packaging, labelling and after-sales service.

4.3.1 Packaging

Again standardisation versus adaptation is the major question. A problem might be the different sizes required in different countries. There are three aspects of packaging.

(a) **Protection**. Packaging may have to be adapted/modified if climate, handling facilities, time spent in distribution chain or usage rate and so on vary.

(b) **Promotion**. Packaging will be adapted if package size, cost of packaging, colour preference, legal constraints, literacy, reputation/recognition varies from market to market.

(c) **Recyclability**. An EU directive has come into force in all European countries requiring a reduction in packaging use, greater re-use and recycling. This has had a significant impact on pack design and distribution channel management for some companies.

Global case study

The market research agency, Research Business International, was commissioned to carry out overseas work for its client, Gillette. Gillette wanted to know more about the market for its razor products in Africa. The agency discovered that in Nigeria, razors were used mainly for skinning animals, not for shaving. Gillette responded by developing a special holder which facilitated this product use.

4.3.2 Labelling

Labelling is an example of mandatory modification required by government regulations.

(a) This usually concerns listing contents or use of appropriate language or languages

(b) Some countries have strict laws regarding the description of the contents of each package

(c) Eco-labelling is now becoming important in some markets informing consumers of the environmental aspects of the product; for example, the energy usage over the lifetime of refrigerators

4.3.3 Servicing

This is an increasingly important part of the augmented product (particularly in the developed economies) and is of great importance in global marketing. If the availability or quality of servicing is doubtful, consumers may choose to buy domestic products.

The service problem is a complex one for the exporter. It involves decisions about facilities, personnel and training. Should they use distributors which would involve training foreign staff and sending out HQ personnel to monitor or should they operate a direct servicing policy in which case they would fly out maintenance staff when required? The appropriate decision varies with the technical sophistication and value of the product.

5 Managing products across borders

5.1 The product life cycle

Many marketing mistakes have been made because firms have failed to take into account the fact that in different countries a product may be at different stages in its product life cycle. You should be thoroughly familiar with this concept.

Marketing principles tell us that products, prices, marketing communications and channels of distribution need to be adapted as a product 'ages' during its life cycle. The marketing mix programme for a new product should be fundamentally different from the mix programme for a mature product.

The product life cycle is relevant to global marketing management. Traditionally many organisations have tended only to operate at home as long as performance there was satisfactory. Then, when domestic performance declined, they tried to close the gap by exporting.

But this is possible only if there are different product life cycle patterns in different countries. In the diagram below, the product is in the decline stage in the home market, in the growth stage in country X, and in the introduction stage in country Y.

Product life cycles in different countries

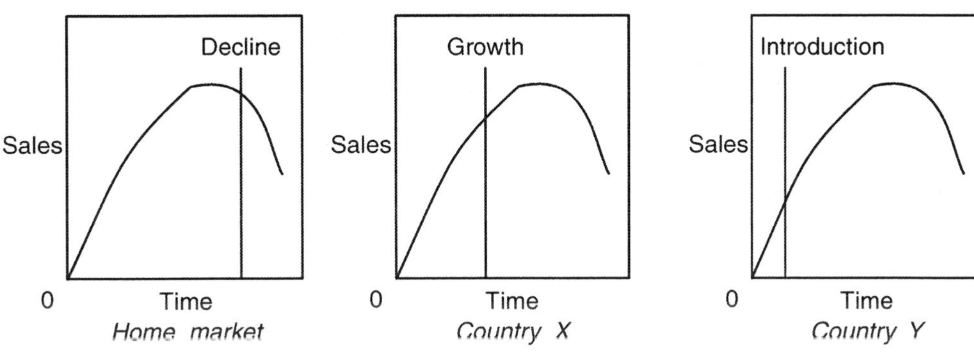

This approach has been very convenient. Organisations were simply able to classify markets according to their economic development and launch declining products in rapid succession into countries with progressively less market development.

Nowadays, however, this type of strategy is less feasible, although not entirely impossible. The revolution in communications among countries during recent years has narrowed the time gap between when saturation occurs in the home market and the last overseas market entered. Hence, the total duration of the profit life cycle, aggregated across all the firm's markets, has shortened. For some products the profit life cycle pattern is exactly the same for home sales as for some/most/all overseas markets.

As a result of these developments, an organisation following a global marketing strategy must consider many markets simultaneously, with a view to implementing a global introduction.

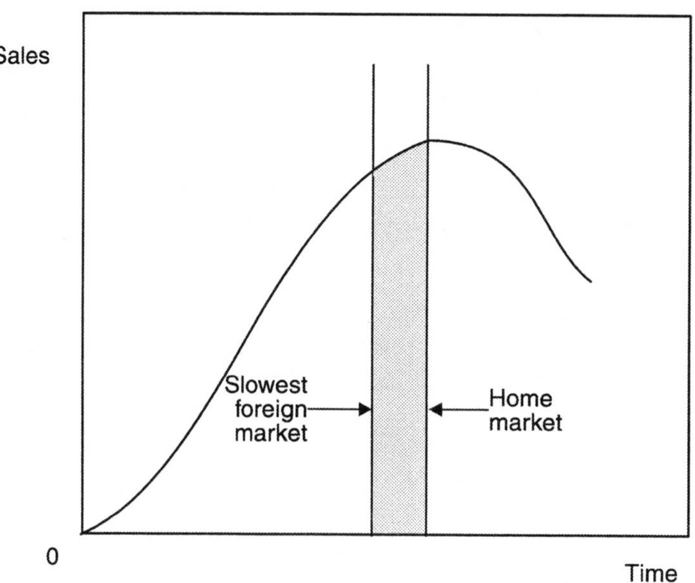

This is necessary to ensure that the product is launched in all potential markets before rivals have time to pre-empt the organisation and to ensure that introduction everywhere coincides with the most appropriate demand conditions.

However, not all organisations operating globally are global corporations and it is therefore important not to ignore the model altogether. The concept is still valid, for example, for an organisation that is not in a fast-changing market.

Because life cycles are generally shortening there is an increased need for a well-developed policy of new product development or repositioning for new markets in order to replace or extend the life cycle.

5.2 The International Trade Life Cycle (ITLC)

Definition

The **international trade life cycle** (ITLC) is the adaptation of the product life model to international conditions.

The ITLC is used in developing long term product strategy. It postulates that many products pass through a cycle during which high income, mass consumption countries are initially exporters but subsequently lose their export markets and ultimately become importers of the product.

From the perspective of the 'initiator high income country' the pattern of development is as follows.

(a) **Phase 1. The product is developed in the high income country.** There are two main reason for this.

    (i) High income countries provide the greatest demand potential.

    (ii) It is expedient to locate production close to the market during the early period so that the firm can react quickly in modifying the product according to customer preferences.

(b) **Phase 2. Foreign production starts.** Organisations in the innovator's export markets start to produce the product domestically.

(c) **Phase 3. Overseas producers compete in export markets.** The costs of the producers in the innovator's export markets begin to fall as they gain economies of scale and experience. These organisations now start to compete in third party export markets.

(d)   **Phase 4. Overseas producers compete in the organisation's domestic market.** The producers in the innovator's export markets become so competitive that they start to compete with the innovator in the innovator's domestic market. The cycle is now complete.

The cycle may well, however continue as organisations in less developed countries enter and ultimately take over the market. The extent and speed of the process depends largely on the product's technical sophistication.

The major significance of the ITLC concept is in long term strategic planning.

(a)   An organisation developing a new product in a high income country should try to penetrate foreign markets as quickly as possible so as to maximise early returns before lower cost organisations located elsewhere enter the market.

(b)   Moreover, the innovator would do well to consider the benefits of establishing production capacity overseas as early as is technically feasible.

5.3 Product portfolio analysis

The use of portfolio approaches in global product management centres around such concepts as the BCG Matrix and the GE Marketing Attractiveness Matrix.

The dimensions of the analysis increase when applied to the organisation's global markets as the competitive positions differ from one market to another.

For these reason the BCG Matrix, for example might be based on one product or one brand in different country markets. This then provides a basis for analysing the current global market portfolio, assessing competitors' product/market strengths and forecasting the likely development of future portfolios both for itself and its competitors.

The portfolio approach to strategic analysis (BCG matrix)

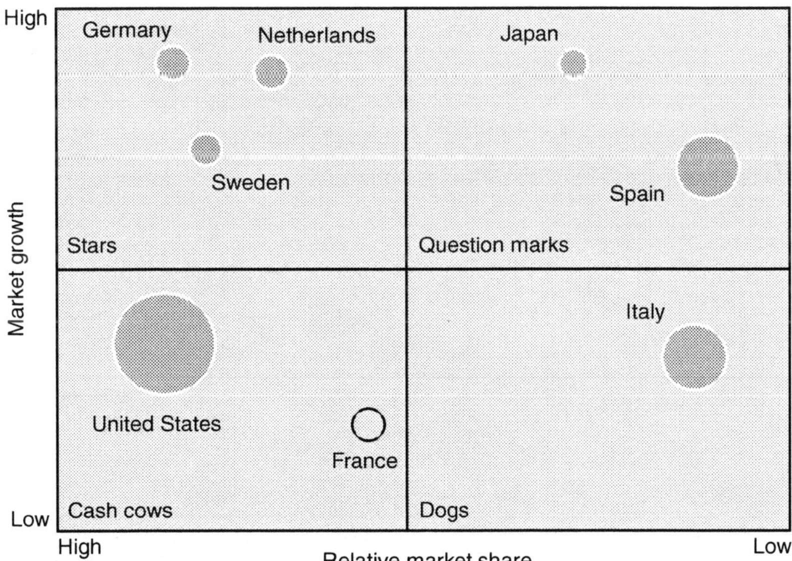

Activity 2

Consider the challenges an organisation might face in marketing products and/or services from:

(a)   A developing country to a developed country
(b)   A developed to a less developed country

How might these challenges be overcome? Illustrate your answer with reference to specific examples.

5.4 Product introduction and elimination

The increased pace of activity has consequences for product management at both ends of the product life cycle.

Too many product introductions in too many countries can risk overburdening the organisation's marketing resources.

Elimination (divestment) should be part of a procedure for product portfolio analysis, where for each country a periodic review of product range is undertaken. Factors to be taken into consideration during this review would include the following:

- Current profitability
- Effects of elimination on the sale of other (complementary) products
- After-sales service implications
- Alternative product opportunities in each country
- The effect on sales/profits of product life extension/rejuvenation

All these factors become much more complex for the multinational company with overseas operations. The multinational company may be producing the product under consideration in a number of countries, under a number of different market conditions.

However, the multinational company's range of alternatives to product elimination for a marginal product is greater than for an exporter. A multinational company can export or license or arrange for contract manufacturing as an alternative to direct manufacturing abroad.

6 New product development for global markets

6.1 NPD process

Sometimes global marketers may need to develop new products for a specific overseas market or group of markets. If a product has been developed for one geographical market, taking it to other countries sometimes requires little extra developmental effort.

Developing products for global markets is similar to that for domestic markets:

- Idea generation
- Idea screening
- Concept tasting
- Business analysis
- Product development and testing
- Test marketing
- Launch and commercialisation

New product development that co-ordinates efforts across national markets does tend to lead to better products. For example Unilever has four global research laboratories that develop products for different national markets while also investigating components for global products.

Where the NPD process does differ in global marketing is the increased level of analysis, co-ordination and communication necessary when assessing the product's suitability and feasibility for a number of markets. Particular emphasis must be paced on the information system and wider consultation must take place with country-specific stakeholders to ensure buy-in, ensure the product meets the needs of the customer in each market and is appropriately and accurately positioned.

Other areas of the NPD process which require careful consideration are:

1   Idea generation: ensure ideas worldwide are assessed so duplication is avoided and synergy is optimised by effectively using all resources to generate new ideas.

2   Idea screening: establish rigorous global criteria to test ideas for suitability in all global regions. It might be that certain products are most suitable only in certain regions.

3   Business analysis: involves establishing criteria for potential success to failure of the product and linking the criteria with regions and/or country markets. Make provision for unexpected events, competitive situation and so on that might have an impact on the business case for that region or market.

4   Product development: ensuring all relevant functions are involved in the process, selecting the most appropriate resources which are matched most effectively with the region or market.

5   Market testing: ensuring the test area is representative of the intended region or market with due regard for distribution and potential competitor response.

6   Launch: can be sequential or simultaneous but consider the competitor actions in each market to ensure that adequate resources are deployed.

A new product to the market can have several degrees of newness from being an entirely new invention through to it being an existing product now available in existing markets. The risk of market failure increases with the newness of the product – the greater the newness the more the need for thorough analysis and consultation in order to mitigate against the risk.

Activity 3

Investigate and consider the ways in which the organisation you are investigating can use the NPD process effectively to enhance and improve its ambitions to become a global brand.

6.2 Success of new products

The success of new products in the global environment depends on a number of factors:

(a)   It is important to have an appropriate organisational structure. A global division, responsive to global rather than purely domestic marketing concerns, is far more likely to introduce new products overseas successfully.

(b)   There should be a commitment to market research. As we have seen, global market research is more complex than domestic market research.

(c)   Sources of idea generation should be as wide as possible: customers, intermediaries, competitors, research and development, sales staff etc.

(d)   The new product development process should be implemented for each country, as far as is practicable.

6.3 Failure of new products

The failure of new products in the global environment depends on a number of factors:

(a)   Tariff barriers and non-tariff barriers
(b)   Local competitor subsidies
(c)   Cultural insensitivity
(d)   Poor planning
(e)   Poor timing
(f)   Lack of a USP in global markets
(g)   Product deficiencies in the market
(h)   Misguided enthusiasm of senior management

Designing a new version of one product for each market is prohibitively expensive, but a company may want more than one product covering all markets. Nissan has been a pioneer in finding the right balance. It reduced the number of different chassis designs from 40 to 8, for cars destined for 75 markets. The development of new products for specific markets is especially relevant to developing nations, although if the market potential is not high a modification to an existing product may be more commercially viable.

Furthermore, there is increasing evidence of time-based competition. In other words many firms are reducing the time spent to get new products researched, designed and launched.

- It wrong foots competitors (eg early mover advantages)
- It enables the firm to get the maximum return from patents
- It might enable the firm to set industry standards for new products
- It enables a premium or skimming pricing strategy

Microsoft launched its XBox games console in March 2002 to 16 European territories, Australia and New Zealand. This followed launches in the US and Japan. At GBP299, many regarded it as too expensive to challenge Sony's PS2. The London launch was hosted by Richard Branson. The games titles reflected a US bias, with no soccer title or Formula One game for European fans. However, the games 'Halo' and 'Project Gotham Racing' proved a hit with 'Gaming 'reviewers.

Speed in NPD can be facilitated by co-ordination between marketing and R&D throughout the design process. There are many good reasons why R&D should be more closely co-ordinated with marketing.

(a)   If the firm operates the marketing concept, then the 'identification of customer needs' should be a vital input to new product developments.

(b)   The R & D department might identify possible changes to product specifications so that a variety of marketing mixes can be tried out and screened.

Other measures to speed NPD:

(a)   Parallel engineering (ie different aspects of the design are carried out simultaneously, rather than being shuffled to and fro in a sequence)

7   Global branding decisions

Definition

A **brand** is a name, term, sign, symbol or design intended to identify the product or service of a seller and to differentiate it from those of competitors. It is a particular make of a product form.

7.1 To brand or not to brand?

The basic advantages of branding include the following.

(a) Branding facilitates memory recall, thus contributing to self selection and improving customer loyalty.

(b) In many cultures branding is preferred, particularly in the distribution channel.

(c) Branding is a way of obtaining legal protection for product features.

(d) It helps with market segmentation.

(e) It helps build a strong and positive corporate image, especially if the brand name used is the company name (eg Kelloggs, Heinz). It is not so important if the company name is not used (eg Procter & Gamble).

(f) Branding makes it easier to link advertising to other marketing communications programmes.

(g) Display space is more easily obtained and point of sale promotions are more practicable.

(h) If branding is successful, other associated products can be introduced.

(i) The need for expensive personal selling may be reduced.

7.2 Types of brand

There are four main choices of brand.

(a) **Individual brand name.** This is the option chosen by Procter & Gamble for example, who even have different brand names within the same product line, eg Bold, Tide. The main advantage of individual product branding is that an unsuccessful brand does not adversely affect the firm's other products, nor the firm's reputation generally.

(b) **Blanket family brand name** for all products, eg Hoover. This has the advantage of enabling the global organisation to introduce new products quickly and successfully. Also the cost of introducing the new product in terms of name research and awareness advertising will be reduced (eg Honda lawn mowers).

(c) **Separate family names** for different product divisions. Alcoholic drinks fall into this category. Most brands of spirits for example are owned by one of a small group of firms.

(d) **The company trade name combined with an individual product name**, eg Kellogg's Corn Flakes, Dyson Turbo, Guinness Malta. This option both legitimises (because of the company name) and individualises (the individual product name).

7.3 Global or local brand?

The key differences between a standardised global brand approach and an approach based upon identifying and exploiting global marketing opportunities are as described below.

(a) **Standardised global brand approach**

  (i) A standardised product offering to market segments which have exactly similar needs across cultures

  (ii) A common approach to the marketing mix and one that is as nearly standardised as possible, given language differences

(b) **Global marketing opportunities**

  (i) A recognition that the resources of the organisation may be adapted to fulfil marketing opportunities in different ways, taking into account local needs and preferences but on a global basis

(ii)   A willingness to sub-optimise the benefits of having a single global brand in order to optimise the benefits of meeting specific needs more closely

(iii)   A global brand is most appropriate when a product has a good reputation

It is possible to move from the latter approach to a global brand approach as demonstrated by the Mars Corporation with their Snickers brand. In the UK market, the biggest 'candy market' in Europe, Mars had decided to use the brand name Marathon for the chocolate bar known as Snickers in the US and elsewhere around the world. Reportedly, this was done to avoid confusion with the word knickers. There was a very distinctive brand identity in the UK to the extent that the company would sponsor the London Marathon and other sporting events to tie in with the brand name. Competition from Nestlé in Europe persuaded the company that they needed to take up the potential benefits of a standardised global brand approach rather than merely relying on a global marketing approach. They, therefore, changed the name to Snickers in the UK market, at very considerable cost of promotional support.

For the global organisation marketing products which can be branded, there are two further policy decisions to be made:

1   The problem of deciding if and how to protect the company's brands
2   Whether there should be one global brand or different national brands for a product

The major argument in favour of a single global brand is the economies of scale that it produces, both in production and promotion. But whether a global brand is the best policy (or even possible) depends on a number of factors, which address the two basic policy decisions above.

7.4 Legal considerations for branding

Legal constraints may limit the possibilities for a global brand, for instance where the brand name has already been registered in a foreign country.

Protection of the brand name will often be needed, but globally is hard to achieve. In some countries registration is difficult and in others brand imitation and piracy are rife.

Budweiser is a US beer, but a beer with an identical name (though a very different taste) is made in the Czech Republic by Budejovicky Budvar and is sold throughout Europe. Anheuser-Busch, the American owner, has been unable to buy the Czech beer's trademark. The two organisations have come up with a practical solution and have formed an alliance. Anheuser-Bush now distributes the Czech organisation's rival beer in the US market under the Czechvar name, giving it access to the Anheuser-Bush network of 600 independent distributors.

Worse still is the problem of piracy where a well known brand name is counterfeited. It is illegal in most parts of the world but in many countries there is little if any enforcement of the law. Piracy is also a problem for intellectual property.

Kaitiki (1981) has identified five ways in which brand piracy takes place:

*   Outright piracy in which the product is in the same form and sues the same trademark as the original but is false

*   Reverse engineering in which the original product is copied and then undersold

Developing Global Marketing Strategy

- Counterfeiting in which the product has altered in some way but the same trademark appears

- Passing off involving modifying the product but retaining a trademark which is similar in appearance

- Wholesale infringement is the questionable registration of the names of famous brands overseas

The organisation exposed to brand piracy faces a number of strategic options:

- Identify and punish the retailer
- Destroy the production company
- Try to convert the pirates to legitimate business
- Accept the pirates and use to highlight the brand's awareness and cache

According to Stuart Whitwell (2006), Joint managing director of Intangible Business some may see black and grey market counterfeiting as advantageous:

- There is no cannibalisation as the people who generally buy fake products are not in a position to buy the original – as soon as their economic condition changes they will run to the front of the queue to buy the original.

- Awareness of the brand rockets in markets in which the authentic product operates as more consumers have access to these products that would otherwise be out of their reach. This increased brand exposure helps drive sales of authentic products from those who can afford them.

- Understanding of the product's values is also not diluted as the owner of the counterfeit products knows it is just that, a fake, and therefore does not expect the same performance from it. In fact, decisions to purchase the counterfeit products usually reaffirm the brand's values as the recipient buys the article to project the very image the brand is trying to portray through its advertising and promotion.

- This endorsement encourages loyalty, generates awareness and strengthens the brand's values with the owner of the fake as well as everyone with whom they come in contact.

- Another reason brand piracy can help increase a brand's value is it is a good indicator of a brand's strength

- Brands which are not faked are considered too weak to generate consumer demand and are consequently not produced. Consumers know this is the case so their decision to buy an original is, in part, derived from the popularity of their chosen brand in the grey channel

- Another benefit of counterfeiting is that it closes off the competition. High priced branded goods encourage competition to enter the market at slightly lower price points. Counterfeiters then produce branded goods and sell at significantly below the cost of the competition. This means the competition is squeezed out as it has nowhere to go – it is priced out of the top market by the original brand and cannot compete with the counterfeit as their prices are too low

- Counterfeiting can give brands access to new markets. For example, a Thai based manufacturer of branded tinned produce used to have a large presence in China. It pulled out a number of years ago and since then the Chinese competition have copied the brand and developed the market. There now remains a much larger market opportunity for the genuine, original brand to exploit by re-entering the market. Awareness has been maintained, penetration has increased and the competition has been closed off.

However counterfeiting can have the potential of damaging the product's brand image and worse, can actually do physical harm.

For example in Colombia investigators found an illegal operation making more than 20,000 counterfeit tablets a day. They contained harmful ingredients including boric acid, cement, floor wax and paint with high lead levels, all designed to replicate the genuine medication's appearance.

Counterfeit drugs can range from those that have the same efficacy through to those that are made with harmful substances.

It is believed that as many as 60% of drugs sold in African and Latin American nations are counterfeit.

Another problem is collusion between the contract manufacturer and illegitimate sellers. Unilever discovered in China that one of its suppliers was making excess soap which was sold directly to retailers. P&G Chinese suppliers sold empty P&G shampoo bottles to another company, which filled them with counterfeit shampoo.

7.4.1 Inadequate protection

Failure to protect intellectual property rights in the marketplace can lead to loss of rights in potentially profitable markets. Some organisations have found their assets exploited in some countries without license or reimbursement.

Worse they then discover that not only are other organisations producing and selling their products or suing their trademarks but are also the rightful owners in those countries where they operate. In such situations the original owner has had to buy back its rights or pay royalties for their use.

When McDonald's entered the Japanese market they registered their golden arches trademark. Only after a lengthy and costly court battle was McDonalds' able to regain the exclusive right to use the trademark in Japan.

Many organisations fail to take adequate steps to legally protect their intellectual property. Some countries do not follow the common-law principle that ownership in one country does not necessarily mean ownership in another.

7.4.2 International conventions

There are two major international conventions that most countries participate in for the mutual recognition and protection of intellectual property rights.

(1)   **The Paris Convention for the Protection of Industrial Property**

The Paris Convention for the Protection of Industrial Property, signed in Paris, France, on March 20, 1883, was one of the first intellectual property treaties. As a result of this treaty, intellectual property, including patents, of any contracting state are accessible to the nationals of other states party to the Convention.

It provides that an applicant from one contracting State shall be able to use its first filing date (in one of the contracting State) as the effective filing date in another contracting State, provided that the applicant files another application within 6 months (for industrial designs and trademarks) or 12 months (for patents and utility models) from the first filing.

The Convention now has 173 contracting member countries, which makes it one of the most widely adopted treaties worldwide.

(2)   **The Madrid Agreement**

The system for the international registration of marks, known as the Madrid system, is the primary international system for facilitating the registration of trademarks in multiple jurisdictions around the world.

The Madrid system provides a centrally administered system of obtaining a bundle of trademark registrations in separate jurisdictions. Registration through the Madrid system does not create an 'international' registration, as in the case of the European CTM system, rather it creates a bundle of national rights, able to be administered centrally. Madrid provides a mechanism for obtaining trademark protection in many countries around the world, which is more effective than seeking protection separately in each individual country or jurisdiction of interest.

Madrid now permits the filing, registration and maintenance of trade mark rights in more than one jurisdiction, provided that the target jurisdiction is a party to the system. The Madrid system is administered by the International Bureau of the World Intellectual Property Organisation (WIPO) in Geneva, Switzerland.

One disadvantage of the Madrid system is that any refusal, withdrawal or cancellation of the basic application or basic registration within five years of the registration date of the international registration will lead to the refusal, withdrawal or cancellation of the international registration to the same extent.

For example, if a basic application covers 'clothing, headgear and footwear', and 'headgear' is then deleted from the basic application (for whatever reason), 'headgear' will also be deleted from the international application. Therefore, the protection afforded by the international registration in each designated member jurisdiction will only extend to 'clothing and footwear'. If the basic application is rejected as a whole, the international registration would also be totally refused.

The process of attacking the basic application or basic registration for this purpose is generally known as 'central attack'. Under the Madrid Protocol, the effects of a successful central attack can be mitigated by transforming the international registration into a series of applications in each jurisdiction designated by the international registration, a process known as 'transformation'. Although transformation is an expensive option of last resort, the resulting applications will receive the registration date of the international registration as their filing date.

In 1997, less than half of a percent of international registrations were cancelled as a result of central attack.

The costs savings which usually result from using the Madrid system may be negated by the requirement to use local agents in the applicable jurisdiction if any problems arise.

 **Global case study**

Some brands do embrace the counterfeit market rather than seeing it as a threat. Georgio Armani, on a recent trip to Shanghai, purchased a fake Armani watch. He said *"It was an identical copy of an Emporio Armani watch... it's flattering to be copied. If you are copied, you are doing the right thing."* Although this was a publicity stunt it does highlight the fact that consumers of fake merchandise are polar opposites to consumers of the authentic and therefore pose no significant threat to the brand owner. Another such example of an opportunity that was taken was with Spanish brandy Fundador in the Philippines. Counterfeit Fundador was rife of, it was an endemic way of life in virtually all Philippines's communities. Children would collect the labels and bottles of the original and sell them to counterfeit production plants who would refill the bottles with a counterfeit product to sell back to the Philippine communities. This eventually offered the brand owner the opportunity to convert them over to the genuine product by developing an extremely generous advertising promotion. Prizes such as Harley Davidsons and Pajero cars were offered on the bottles' labels which the counterfeiter's could not either compete with or copy because the offers were changed every six weeks. Sales of the genuine product quadrupled in three years with this strategy.

http://www.intangiblebusiness.com/Brand-services/Marketing-services/News/Brand-piracy-faking-it-can-be-good~290.html

7.5 Cultural aspects of branding

Even if an organisation has no legal difficulties with branding globally, there may be cultural problems, eg unpronounceable names or names with other meanings. There are many examples of problems in global branding. The Internet again has cause blurring in the global/local issue for brands.

7.6 Other marketing considerations

Many other influences affect the global branding decision, including:

(a)   Differences between the organisation's major brand and its secondary brands. The major brand is more likely to be branded globally than secondary brands.

(b)   The importance of brand to the product sale. Where price, for example, is a more important factor, then it may not be worth the heavy expenditure needed to establish and maintain a global brand in each country; a series of national brands may be more effective.

(c)   The problem of how to brand a product arising from acquisition or joint venture. Should the multinational company keep the name it has acquired, for example?

In addition concern for global environmental issues is becoming increasingly important in many countries and has implications for product and service policies.

1 Explore global product and services marketing

- Global products and services can be defined in the same way as we distinguish between any product or service, the cultural nature and implementation of marketing strategy however is likely to differ.

- Services possess the following characteristics: intangibility, inseparability, variability/heterogeneity and perishability.

- A **product** is anything that can be offered to a market for attention, acquisition, use or consumption that might satisfy a want or need. It includes physical objects, services, places, organisations, ideas, and people.

2 Consider the balance between standardisation and adaptation

- In global marketing, the choice is between standardising the product in all markets to reap the advantages of scale economies in manufacture and, on the other hand, adaptation which gives the advantages of flexible response to local market conditions. Adaptation may be unavoidable in some circumstances (eg if different countries have different legal standards for safety).

- Packaging, labelling and even branding must sometimes be sensitive to the local conditions

3 Evaluate global product policy

- Organisations may have a single global brand, because of the economies of scale that may be realised, but global branding may not be the best strategy and there are a number of other strategies that can be deployed.

4 Assess the management of products across borders

- A product's stage on the product life cycle may vary from country to country, and different marketing strategies will be appropriate in each case.

5 Evaluate new product development for global markets

- Time based competition is becoming increasingly important in new product development. Product development may be co-ordinated across several international markets.

6 Consider global branding decisions

- Different product types may require different launch strategies in different markets. Understanding that an established product in one market may be considered an innovation in another is critical in the global planning and development of products and services.

1   The answer to this activity will depend largely on the service sector you are considering. You will need to consider the extended marketing mix including includes people, processes and physical evidence but there are also issues of overall service decision, quality and delivery and management.

2   The answer to this will largely depend upon the organisation you have chosen and the countries you are considering. You could consider applying the International Trade Life Cycle and proposing recommended strategies that overcome the challenges to market entry.

3   The answer will largely depend upon your organisation, its attitude to New Product Development and the extent to which it already has processes in place.

Cateora, P. and Graham, J., (2009). *International Marketing*. 14th edition. London: McGraw Hill.

Doole, I. and Lowe, R., (2008). *International Marketing Strategy Analysis, Development and Implementation*. 5th edition. London: Thomson Learning.

Hollensen, S., (2007). *Global Marketing*. 4th edition. London: Prentice Hall.

Kaitiki, S., (1981). 'How multinationals cope with international trade mark forgery'. *Journal of International Marketing*. Vol 1 Issue 2, pp. 69-80.

Keegan, W. J., (1995). *Multinational Marketing Management*. London: Prentice Hall.

Whitwell, S., (2006). 'Brand Piracy: faking it can be good'. Intangible Business.

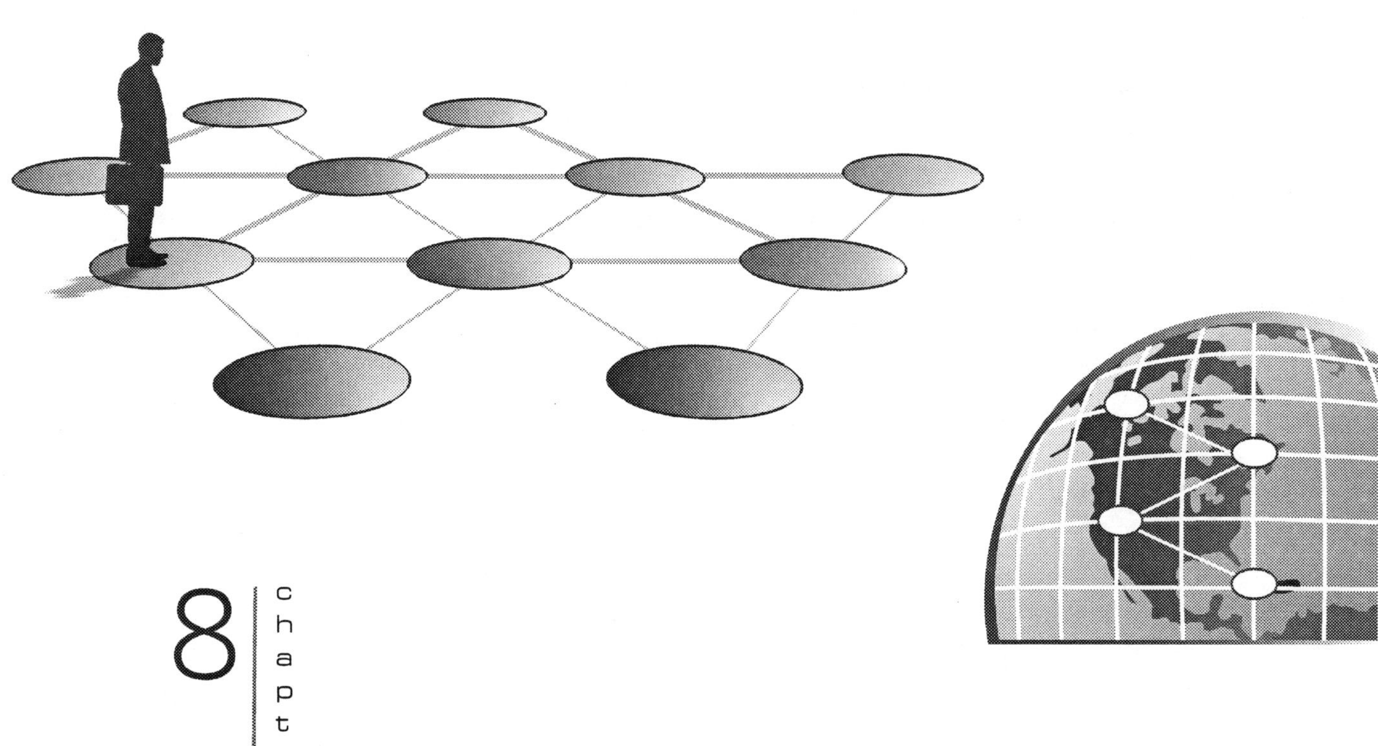

8 | chapter

Global pricing decisions

Setting the right price for a product or service can result in success or failure and of all the tasks facing the global marketer determining what price to charge is one of the most difficult. It is further complicated when an organisation moves beyond domestic marketing to global marketing and must be recognised as an important consideration at a strategic level. As globalisation continues and competition intensifies price becomes increasingly important as a competitive weapon in the marketing mix. The challenge for the global marketer is to set and control a pricing strategy in multiple markets giving due consideration to the environmental variables in each country or region in which the organisation operates or seeks to operate.

Contents

Chapter learning outcomes

In this chapter you will cover the following:

- Examine the role of pricing
- Evaluate the factors influencing global pricing decisions
- Consider global pricing approaches
- Identify terms of sale
- Assess methods of payment
- Consider the issues of countertrade and leasing

1 Setting the scene

At the best of times, pricing is a difficult decision for global marketers, as it is what the consumer is actually giving in response for the satisfactions offered by the marketing mix. In this session we focus on both the internal and external factors that affect global pricing decisions, the role that pricing plays in developing strategies that meet corporate objectives, and the relationship between pricing and other aspects of the organisation's activities. We also explore the specific problems associated with developing global pricing strategies, the financial issues in managing risk in pricing, payment and delivery terms as well as a discussion of countertrade as a pricing tool.

2 The role of pricing

Pricing is the only mix decision that produces revenue. The other elements involve costs. It is often the major mix decision yet its importance can be overstated. For many products, other mix aspects are more important, such as product quality, augmented product features, distribution and so on.

Many organisations do not handle pricing well. Common mistakes include the following.

- Pricing is too cost-orientated
- Pricing is not revised often enough to reflect changing market conditions
- Pricing is decided in isolation rather than as part of an integrated marketing plan
- Pricing is not flexible enough to meet the needs of different segments or countries

Flexible pricing is necessary to cover the following situations.

(a)   When a new product is launched.
(b)   When the company wants to initiate a price change.

- Cost increases
- A decision to sell as a loss leader
- A sales promotion
- A change in the product life cycle stage
- A change in discount policy
- A decision to reposition the product

(c)   As a response to price change by competitors.
(d)   When the company wants to decide on a price policy for an entire product line.

Successful global marketing involves an analysis of the extent to which prices should be adapted to meet the different environmental and competitive situations in the company's markets.

3 The factors influencing global pricing decisions

The pricing policy deployed in a global context is both an important strategic and tactical competitive weapon that is highly controllable and is generally inexpensive to change or implement. It is therefore important that pricing strategies are integrated with other elements of the global marketing mix.

3.1 Key global pricing factors

According to Hollensen (2007) factors affecting global pricing can be broken down into two main groups, internal factors and external factors. The framework below presents a general framework for global pricing decisions.

Global pricing framework

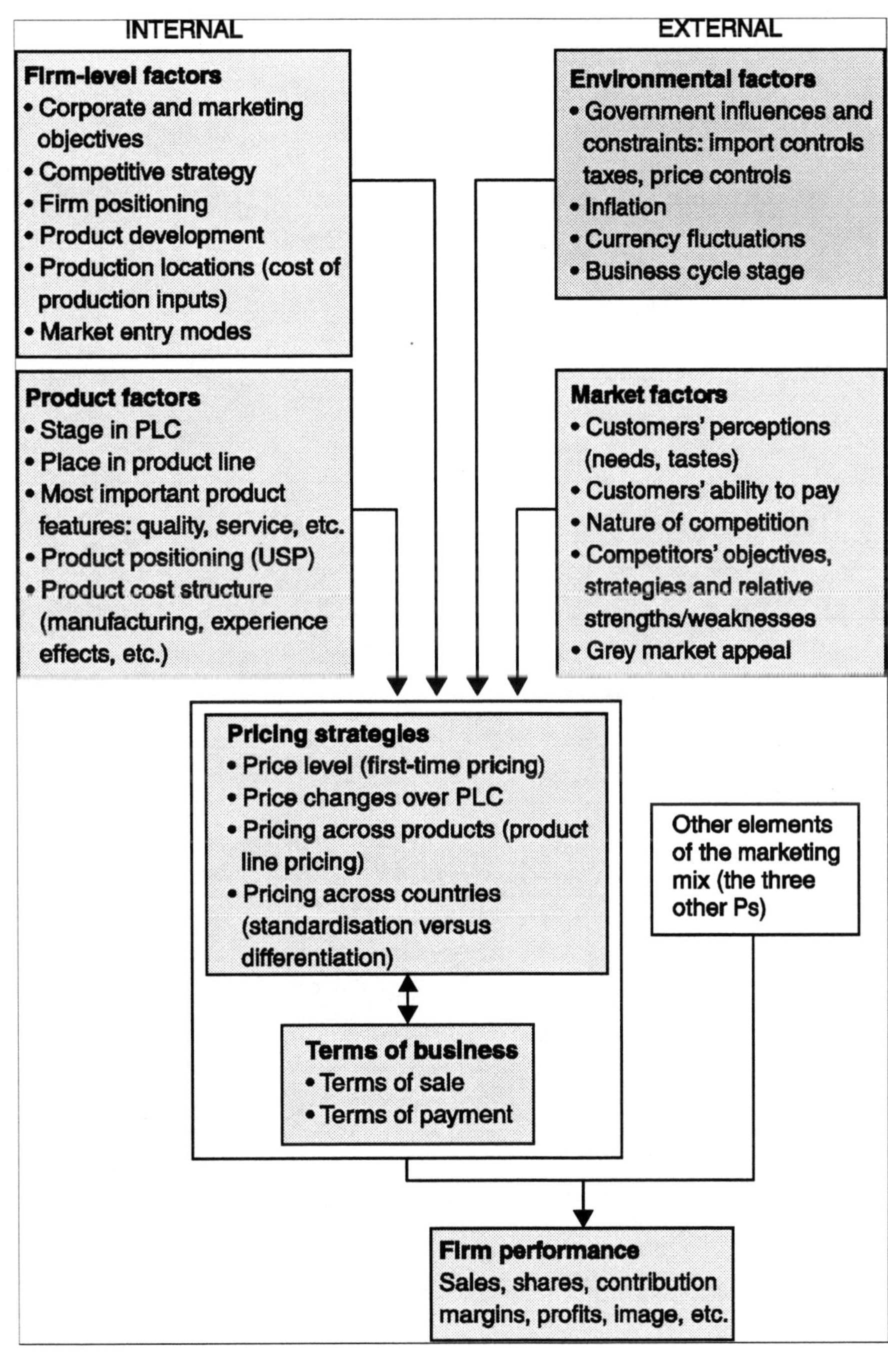

Factors	Comments
Organisational factors	• Influenced by organisational vision and mission • Market share objectives • Preparedness to wait for profits from a new markets • Support and attitude of corporate investors • Choice of market entry mode
Product and service factors	• Uniqueness of the product • Extent to which the product or service has had to be modified • Level to which the market requires service around the core product • Extension of the distribution chain in delivering the product or service • Frequency of purchase – more frequently purchased products are likely to be more price sensitive • Degree of comparability with existing products in the market • Degree of fashion or status of the product within the country
Environmental factors	• External to the organisation such as government control of exports and imports, tariff levels and so on • Fluctuations in the exchange rate
Market factors	• Purchasing power of the customers in the markets • The more competition that exists in the market the closer the prices must be and the greater the influence of costs in determining prices • Any price differentials must be justified in the mind of the customers on the basis of perceived value • The role of price in differentiating products within a market within a category can be used an offensive strategy

3.2 Price escalation

Definition

All cost factors in the distribution channel add up and lead to **price escalation**. The longer the distribution channel the higher the final price in the foreign market.

Unless some of the costs that create price escalation can be reduced, the global marketer is faced with a price that may confine sales to a very limited segment of the total potential market. Once price escalation is set in motion it can spiral very quickly. The following options are available to counter price escalation:

3.2.1 Rationalising the distribution process

• Reduce the number of links in the distribution channel, either by doing more in-house or by circumventing channel members.

3.2.2 Lowering the export price from the factory

• Reduces the effect of all mark-ups along the distribution chain.

3.2.3 Establishing local production of the product

- Eliminates some of the cost out of the distribution chain. Local prices can best be achieved with local overheads.

3.2.4 Pressurising channel members to accept lower profit margins

- More appropriate if the intermediaries are dependent on the manufacturer for much of their turnover.

Price escalation (examples)

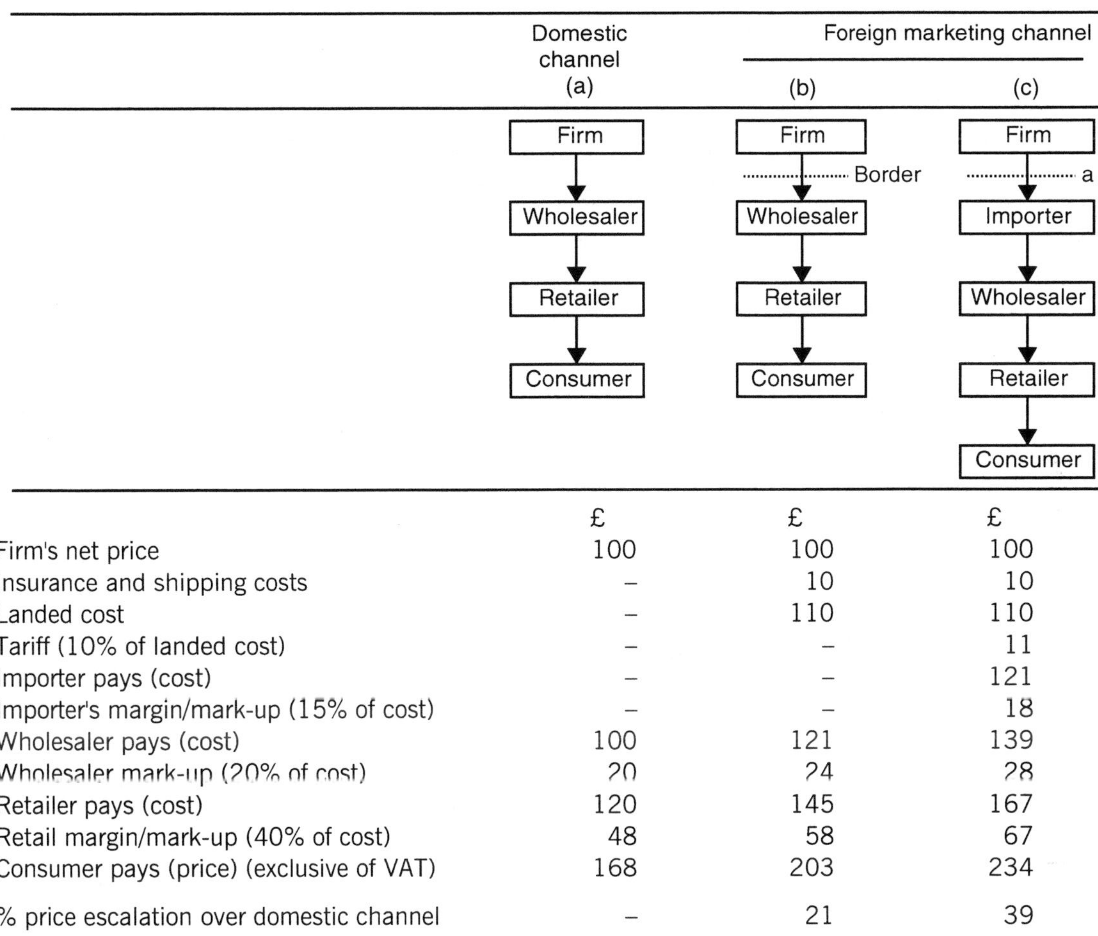

	Domestic channel (a)	Foreign marketing channel (b)	(c)
	£	£	£
Firm's net price	100	100	100
Insurance and shipping costs	–	10	10
Landed cost	–	110	110
Tariff (10% of landed cost)	–	–	11
Importer pays (cost)	–	–	121
Importer's margin/mark-up (15% of cost)	–	–	18
Wholesaler pays (cost)	100	121	139
Wholesaler mark-up (20% of cost)	20	24	28
Retailer pays (cost)	120	145	167
Retail margin/mark-up (40% of cost)	48	58	67
Consumer pays (price) (exclusive of VAT)	168	203	234
% price escalation over domestic channel	–	21	39

Global case study

Over the past few years, rising commodity and energy prices caused consistent price escalation in many consumer packaged goods and other products sold to consumers through grocery, mass merchants and other retail channels. The manufacturers, used to little leverage over the past many years due to the growing clout of large retail chains, blamed the rising price of oil (transportation, plastics), corn, wheat, and anything else they could cite as increasing underlying cost drivers. This price stance, however, has not gone over well with retailers, which argue that materials cost changes should lead to downward price changes now just as they caused increases in 2005-2008. Those tensions boiled over in Europe this week, as Brussels-based Delhaize SA, which also owns the Food Lion chain in the US, pulled some 300 products of CPG giant Unilever that it believes are priced too high off of the shelves of its 775 stores in Belgium this week. The stores stock some 500 Unilever products in total, but about 40% were not pulled because the prices were not deemed excessive. Delhaize says Unilever has been trying to force its full product assortment into its stores, including some that the grocer says it would prefer not to stock. If the supermarket doesn't buy the whole range of products, Delhaize says, Unilever has threatened to raise prices by an average of 30% for the remaining items, according to the Wall Street Journal. The stand-off comes as there is data showing more consumers are switching to cheaper, private-label brands in many cases. Meanwhile, some stock industry analysts have warned grocery chains may be under pressure as the economy causes a "price war" that will eat into revenues and profits, making the price negotiations even more critical.

http://www.scdigest.com/assets/newsviews/09-02-11-1.php?cid=2250&ctype=content

Discussion

Study buddy

Share your thoughts with your Study buddy as to the major causes of price escalation for the organisations you are investigating. Suggest possible courses of action to deal with this problem. Together then share with the wider cohort through the discussion forum the conclusions you have reached.

4 Global pricing approaches

4.1 Developing pricing strategies

4.1.1 Standardisation versus differentiation

In the global context the first question that needs to be considered is to what extent prices should be standardised or adapted. There are essentially two opposing forces:

1   Achieving similar positioning in different markets by adopting largely standardised pricing
2   Maximising profitability by adapting pricing to different market conditions

There are three approaches to global pricing strategies:

- **Standardised, or ethnocentric pricing**

    A single price is charged to recover costs and earn the return. As this is translated at local exchange rates, this can lead to price volatility in local terms and a lower volume of sales than would be possible. The demand curve in the overseas market may differ, and using a fixed standard price may not result in a maximised marginal revenue in the overseas market.

- **Adaptation or polycentric pricing.**

    Each local subsidiary sets its own prices. There is no co-ordination. Headquarters has no control, and 'grey markets' develop.

- **Geocentric pricing**

    Aims to have a global pricing strategy, but in the short term at least (eg introducing a new product) local subsidiaries may have some autonomy.

Where products or services are both produced and sold in the overseas market the extent of pricing control exercised from outside that country will depend on the organisation's structure.

- **Ethnocentric approach**: group headquarters dictates its subsidiary's pricing decisions totally.

- **Polycentric approach**: the subsidiaries are autonomous. This is fine if it manufactures its products locally, as costs and revenues will be in the same currency.

- **Geocentric approach**: control varies depending on the situation, perhaps the best of both worlds.

Consider organisations that operate in the market sector within which your orgainsation operates and identify the pricing approaches they are taking. At the same time consider the approach that BPP Learning Media and GMN have taken and outline the rationale for their approaches.

4.2 Pricing strategies

In addition to the factors discussed earlier an organisation's pricing strategy will also be affected by its attitude to global marketing and its sensitivity to market and environmental conditions.

The most common pricing objectives are:

4.2.1 Rate of return

Cost-oriented organisations set prices to achieve a specific ROI and may quote the same for both domestic and global markets.

4.2.2 Market stabilisation

An organisation may choose not to provoke a market leader, so that market shares remain significantly unchanged.

4.2.3 Demand-led pricing

Prices are adjusted according to assessment of demand. Demand led pricing enables marketers to set prices according to the customer's ability and willingness to pay. These demand levels may vary from one country to another, within the same country among different segments, or even within the same segment over time.

This is the true marketing based approach and puts customers' needs and preferences at the heart of the pricing decision. Demand led prices are most prevalent in branded consumer goods but are increasingly possible in many industrial goods markets.

4.2.4 Cost based pricing

Total cost is often the basis for pricing. This is a sensible policy in markets where price is the only or most important factor in the purchase decision and where there is little differentiation among product offerings, both at a formal and augmented level, (eg industrial nuts, bolts and screws).

However, research tends to suggest that too many firms adopt cost based pricing for all their products and in all their markets when other approaches would be more effective in achieving their objectives. For example, even in industrial product markets demand intensities can vary between countries, which gives opportunities for price differentiation.

4.2.5 Competition and pricing

There are two aspects of competition-based pricing relevant in the global context.

Where there is almost perfect competition individual suppliers of a product have no control over the price they charge. This is the case with commodity prices such as those for tea and coffee. Here current world market prices are known to customers and any change is established as a result of interaction among a large number of buyers and sellers.

Where an organisation can base its price levels in relation to its competitors in order to achieve certain objectives the price charged must be consistent with those objectives.

The price/quality diagram below illustrates the competitive price strategies open to an organisation.

		Relative price		
		High	*Medium*	*Low*
product	*High*	1 Premium pricing strategy	2 Penetration pricing strategy	3 Super bargain strategy
quality	*Medium*	4 Overpricing strategy	5 Average pricing strategy	6 Bargain pricing strategy
	Low	7 Hit & run pricing strategy	8 Shoddy goods pricing strategy	9 Cheap goods strategy

Some examples of the model's application in the international context would be as follows.

(a) An organisation might adopt Strategy 3 (at least in the short term) in one country in order to penetrate a difficult competitive market while adopting Strategy 1 in another where it already has a good reputation and fewer real competitors.

(b) A organisation might adopt Strategy 9 (cheap goods strategy) in a less developed country because of low disposable income levels while producing a better quality product at a higher price in a market where disposable incomes are higher (Strategies 5 or 1).

(c) Any of Strategies 4, 7 or 8 (over pricing, hit and run pricing or shoddy goods pricing) might be adopted in markets where customers are unaware that they could obtain the same quality at a lower price or a higher quality at the same price. In another, more knowledgeable market, a cheap goods strategy (9) might be used.

Kotler states that Strategies 4, 7 and 8 should be avoided by professional marketers. Selling shoddy goods devalues the brand name, and will discourage repeat purchases.

4.2.6 New product pricing

There are three elements in the pricing decision for a new product. These are as follows.

- Getting the product accepted

- Maintaining a market share in the face of competition

- Making a profit from the product - when an organisation launches a new product on to the market, it must decide on a pricing policy which lies between the two extremes of market penetration and market skimming.

Global case study

In the early 1980s, low-cost disposables had become the hot shaving product worldwide, grabbing market share from shaving systems. Many at Gillette were about to throw in the towel on quality to concentrate on price. However, Gillette's leadership stood its ground and continued to invest in quality spring-mounted systems, labelled Flag (Floated Angle Geometry). Sensor went on to far exceed all its goals: 24 million (versus 18 million target) razors sold and 350 million (versus 200 million target) blade cartridges sold in the first year of launch.

http://www.echeat.com/essay.php?t=30238

4.2.7 Penetration pricing

Definition

Market penetration pricing is a policy of low prices when the product is first launched in order to gain sufficient penetration into the market. It is therefore a policy of sacrificing short-run profits in the interests of long-term profits.

The circumstances which favour a penetration policy are as follows.

- The organisation wishes to discourage rivals from entering the market.

- The organisation wishes to shorten the initial period of the product's life cycle, in order to enter the growth and maturity stages as quickly as possible.

- There are significant economies of scale to be achieved from a large output. An organisation might therefore deliberately build excess production capacity and set its prices very low; as demand builds up, the spare capacity will be used up gradually, and unit costs will fall; the organisation might even reduce prices further as unit costs fall. In this way, early year losses will enable the organisation to dominate the market and have the lowest costs.

Global case study

With the current economic turmoil, car sales in the United States and the rest of the world for that matter have been very sluggish. The big three car manufacturers (General Motors, Chrysler and Ford) are teetering on the brink of bankruptcy. Not to say that Toyota has done considerably better than the big three, Toyota has managed to strategically market its products to its target market better than the big three, in this case the Toyota Prius. Toyota's hybrid market has been segmented according to the buyers behaviour, in this case, buyers who are pro environmentalists. Toyota has used the penetration pricing strategy instead of the skim pricing strategy because the demand for hybrid cars is elastic, and the quantity of cars bought will increase as prices decline plus the car itself will gain mass appeal faster. With Toyota's strong balance sheet, Toyota is also able to compete with other car makers by offering competitive financing and lease rates throughout all its 1200 distributors in the United States alone.

http://www.associatedcontent.com/article/1804334/marketing_mix_in_actiontoyota_prius.html

4.2.8 Skimming

Definition

Market skimming pricing involves the following. Some firms set a high initial price to achieve high unit profits, knowing that a certain number of customers will buy at the high price. This is possible where rival firms are not expected to undercut these high prices, where the fixed costs of output are fairly low, so that economies of scale are relatively insignificant and where the customer believes that high prices signify a quality product.

Market skimming involves the following.

- Charging high prices when a product is first launched.

- Spending heavily on advertising and sales promotion to win customers.

- As the product moves into the later stages of its life cycle progressively lower prices will be charged. The profitable 'cream' is thus 'skimmed' off in progressive stages until sales can only be sustained at lower prices.

The aim of market skimming is to gain high unit profits very early on in the product's life.

Conditions which are suitable for such a policy are as follows.

- Where the product is new and different, so that customers are prepared to pay high prices so as to be 'one up' on other people who do not own one. Many new technology items come into this category.

- Where demand elasticity is unknown. It is better to start by charging high prices and then reducing them if the demand for the product turns out to be price elastic than to start by charging low prices and then attempting to raise them substantially when demand turns out to be price inelastic.

- High initial prices might not be profit-maximising in the long run, but they generate high initial cash flows.

- Skimming may also enable the firm to identify different market segments for the product, each prepared to pay progressively lower prices. If product differentiation can be introduced, it may be possible to continue to sell at higher prices to some market segments.

Assessment advice

It is now useful to begin to construct the pricing strategy for the organisation you are investigating. Consider the options in light of current and past performance and evaluate the approach you are recommending for the country you are proposing to enter by considering the advantages and disadvantages of each potential approach.

4.3 Transfer pricing

Definition

The **transfer price** is the price at which goods or services are transferred from one process or department to another or from one member of a group to another. The extent to which costs and profit are covered by the price is a matter of policy. A transfer price may, for example, be based upon marginal cost, full cost, market price or negotiation. Transfer pricing is important in global businesses which source components from many countries from their various operating divisions.

When a multinational firm adopts a decentralised organisational structure, each of its manufacturing units becomes a profit centre. Components, semi finished or finished products may have to be transferred between these manufacturing or assembly units.

It is in this context that the question of transfer pricing arises. If these components or products are sold on the open market they will be sold at arm's length price (market price). Multinational firms must decide whether the transfer price between units of the same organisation should be equal to, higher than or lower than the open market arm's length price. Once again, problems associated with transfer pricing in a domestic marketing situation become more complex in an international context.

4.3.1 Setting the transfer price

Transfer price less than open market price

A multinational organisation will find it beneficial to set its transfer price below the open market price in the following situations.

- If the importing country has a lower rate of profits than the exporting country. In this situation, for the ethnocentric organisation, dividend repatriation is easy. The problem here is that operating revenues and profits are distorted and problems of managerial motivation and appraisal arise.

- If the importing country has high tariff barriers, the impact of those barriers may be lessened by charging a price below open market price and thus improve overall corporate profits.

- As a competitive pricing strategy, it may be possible to penetrate a new, competitive market more quickly by adopting a low initial transfer price (provided the market has high price elasticity).

- If inflation is high in the exporting country it may be possible to transfer funds by high transfer pricing to an economically 'safe' country.

Transfer price greater then open market price

In the following circumstances it may be beneficial to the organisation to set the transfer price above open market price.

- If the importing country taxes profits at a higher rate of tax than the exporting country.

- If dividend repatriation is restricted.

- If tariffs are at a relatively low level in the importing country.

- If there is a fear of expropriation of assets the organisation will wish to hold its cash assets in the most politically 'stable' country.

4.3.2 Problems of transfer pricing

The problem of managerial motivation and evaluation may override a transfer price that has the best economic justification. Tax avoidance strategies are increasingly attracting the attention of governments and their tax authorities. Because of this 'public surveillance' international companies increasingly feel the need to be seen to be playing fair to both domestic and host countries and to unions and other groups. Finally, transfer pricing might, indirectly, be the subject of anti dumping actions.

4.4 Foreign currency pricing

> **Definition**
>
> **Foreign currency pricing** involves pricing your goods in foreign currency, thereby suffering the exchange risk. This risk can be controlled by hedging strategies.

In international trade, the exporter must invoice the buyer in a foreign currency (eg the currency of the buyer's country) or the buyer must pay in foreign currency (eg the currency of the exporter's country).

It is also possible that the currency of payment will be the currency of a third country; for example, a UK firm might sell goods to a buyer in Brazil and ask for payment in US dollars. One problem for importers is therefore the need to obtain foreign currency to make a payment, and for exporters there can be the problem of exchanging foreign currency received for currency of their own country. Banks provide the service to importers and exporters of buying and selling foreign currency.

The cost of imports to the buyer or the value of exports to the seller might be increased or reduced by movements in foreign exchange rates. Although there is a chance of making a profit out of favourable movements in exchange rates, movements in foreign exchange rates introduce a serious element of risk which might deter firms from entering international sales or purchase agreements.

4.4.1 Reducing risk

Match receipts and payments

The foreign exchange risk does not arise for a business that makes payments and earns receipts in the same foreign currency, because payments in the currency can be made out of cash income in the same currency.

For example, if a UK company buys goods from supplier X costing US $10,000, and at the same time the UK company sells goods abroad to customer Y for US $10,000 (in dollars), the company can use the US $10,000 it receives from customer Y to pay the US $10,000 dollars to supplier X. If 'matching' receipts and payments is carried out in this way, the exchange rate between the foreign currency and the company's domestic currency would be irrelevant and exchange risk would be avoided.

Matching currency receipts and payments (or currency receipts and the repayment of currency loans) is only feasible if the international trader has receipts and payments in the same currency to match.

The futures market

A futures contract is a contract for the delivery of a standard package of a standard commodity at a specific point in the future. The 'commodity' may be currency or a commodity in the traditional sense, ranging from aluminium and cattle to wheat, wool and zinc.

To take an example, let's say that the price today for a tonne of cocoa is $500. An international marketer of cocoa knows he will have one tonne of cocoa available for sale in a month's time. He does not know what the price will be, so he may decide to reduce the risk by selling at the current price a contract for a tonne of cocoa to be delivered in one month. He has removed uncertainty. Whatever happens to the price of cocoa over the next month, he knows how much money he will get.

The futures market can therefore help a company by allowing it to plan ahead on the basis of known prices.

Forward markets and hedging

It is up to the international company to protect itself against the swings of currency value. There are various ways that this risk can be controlled by hedging.

There is a forward market in currencies as well as a spot market. In the forward market, currencies are bought and sold for delivery in the future. Spot prices are quoted for immediate delivery.

Suppose a US company knows it is due to receive Danish krone (and will therefore have Danish krone to sell) in two months' time. It will be worried that the value of the krone will fall, meaning it will receive less in sterling terms. So it might sell Danish currency today for delivery in two months' time, locking into a known exchange rate.

Quotation in own currency

Price quotation in own currency is administratively convenient, an important factor for the small firm or the firm where export revenue is a small proportion of total revenue. Effectively the exchange rate risk of variation is borne by the foreign customer.

Quotation in foreign currency

Quotation in foreign currency means that the exporter accepts any exchange risk of fluctuating values and has to deal with it. These risks become even greater the 'softer' a currency is (ie the less easy to convert into other currencies). The exporter might be made vulnerable to a severe risk of loss, unless the exporter engages in hedging.

Advantages of foreign currency pricing are as follows.

(a)   An exporter can gain a competitive edge by quoting in a foreign currency if the importer would prefer his own currency, either to avoid exposure to currency fluctuation or because of exchange control regulations. Foreign currency pricing also makes it easy to relate to retail prices overseas. In addition, constant adjustment to a sterling price list is avoided.

(b)   Foreign currency invoicing can sometimes help exporters to borrow at lower rates of interest than at home.

(c)   The exporter may purchase the foreign currency 'forward' on the exchange markets (ie at a fixed rate some time in the future). The exporter can use that particular rate, which may be better than later 'spot' rates in some cases, or at least the risk of loss is avoided.

Resolving differences between exporter and importer

Despite the advantages of foreign currency pricing noted above, many exporters frequently prefer to quote prices in their home currency. It makes calculation of profits, cash flows and possibly costs easier and avoids short-term exposure to exchange rate risks. But foreign customers normally prefer to receive price quotations in their own currency since it places the risk of exchange rate fluctuation on the exporter and facilitates comparison of prices from a number of foreign sources.

Resolution of this potential conflict depends on several factors.

- The relative strengths of the two parties
- The importance of the contract to either party
- The economic situation in both countries (such as balance of payments position)

Whichever currency is used there are a number of ways of minimising problems caused by foreign currencies:

- The simplest way is to have an appropriate clause in the contract to adjust the price if it fluctuates within agreed limits or to renegotiate if these are exceeded.

- To purchase (or sell) forward as appropriate whereby a fixed rate is agreed for the given sum at a specified date in the future by a bank. The net sum (after commission) received or paid at the end of the period is at least guaranteed.

- In a situation where either or both currencies are unstable a third currency can be quoted, eg the US dollar. A strengthening or weakening in one party's currency would be felt in relation to the dollar and the other party would not be affected. Each party to the transaction thus takes on the risk associated with its own currency. If the dollar changed value, it would be likely to be relative to both currencies and the risk (or benefits) would be shared. The one disadvantage is that there is a double exchange transaction involved. However, this may be more acceptable than other options.

5   Terms of sale

In addition to the obvious manufacturing costs of the exporter there may be significant extra costs in getting the goods to a foreign buyer. Generally, the costs of physical movements are as follows.

- Transport from the manufacturer's premises to the docks
- Loading aboard ship
- Freight charges
- Unloading
- Customs duties
- Transport from the docks to the customer's warehouse

As well as these costs of physical movement there are also insurance charges and possibly additional costs for any delays.

There are a number of internationally accepted standard forms of dividing these costs between buyer and seller. Who pays what, and who is responsible for arranging for the transport, has to be confirmed when the sale contract is agreed.

5.1 INCOTERMS

The standard forms are known as INCOTERMS and have been designated by the International Chamber of Commerce. You will note below that many of the descriptions contain the term 'free'. This indicates the time at which legal title passes from seller to buyer (and thus risk of loss in case of damage, theft and so forth). Among the more common forms of quotation are the following.

5.1.1 EXW: Ex works

The buyer must take delivery at the exporter's factory and pay all the costs of freight, insurance and other expense items to get the goods transported from the supplier's factory to their destination. This represents the minimum obligations for the seller.

5.1.2 FAS: free alongside ship

The seller arranges the following.

- Delivery of the goods alongside the named ship at the port of loading named in the contract

- Payment of all the charges up to delivery of the goods alongside ship, including freight and insurance charges to that point

The buyer is responsible for the following.

- Choosing the carrier to transport the goods abroad and paying the cost of freight from the port of shipment including the cost of loading the goods on board ship (if loading costs are separate from freight charges)

- Arranging insurance and paying insurance from arrival at the dockside onwards

- Arranging and paying for any export licence or export taxes

5.1.3 FOB: free on board

FOB means that the buyer does not have to pay for transporting or insuring the goods from the place where they are originally despatched up to the point when they are taken on board ship. The costs up to this point are borne by the seller/exporter. The place of delivery is the ship's rail.

The seller/exporter:

- Pays for transportation, freight and insurance charges to the named port of shipment (eg 'FOB Los Angeles' would mean that an American supplier would be responsible for sending goods for shipment on board at Los Angeles, and to pay costs up to that point)

- Provides and pays for the export licence

- Pays export taxes

- Delivers the goods on board the ship (or airline flight) that the buyer has specified

- Pays for the cost of loading the goods on board ship (if loading costs are separate from freight charges)

The buyer:

- Nominates the carrier to transport the goods (eg, if the shipping terms for an export consignment from the UK are FOB Stranraer, it is the buyer who specifies the shipping company, sailing date and time)

- Gives the seller the details of the ship and sailing time

- Pays for the carriage from this point (ie freight from that point, including costs of unloading at the place of destination)

- Arranges and pays for insurance of the goods from this point

5.1.4 CFR: cost and freight

With Cost and Freight, the exporter/seller must nominate the carrier to ship the goods abroad, arrange the contract of carriage and pay freight charges. In these respects, CFR differs from FOB.

Though the supplier pays the freight charges to the port of destination, the place of delivery of the goods is the ship's rail when the goods are taken on board. When they are on board, they are the responsibility of the buyer even though the supplier pays freight charges.

The seller:

- Nominates the carrier and so makes the contract of carriage

- Pays for transportation of the goods to the place of shipment and insures the goods up to this point

- Provides and pays for the export licence

- Pays export taxes

- Delivers the goods on board

- Pays for the cost of loading the goods if the loading charge is separate from the freight charge

- Provides the buyer with a clean on board bill of lading

- Pays the cost of freight charges to the named port of destination (eg, 'CFR Rotterdam' would mean that the UK exporter must pay freight charges for delivery to the port of Rotterdam)

- Sends to the buyer advice of the carrier and the shipment date

The buyer:

- Pays for the insurance of the goods from the time they are taken on board, and so bears the risk of loss or damage to the goods

- Pays for unloading costs at the port of destination if these costs are separate from the freight charges

- Pays for any import licence required

- Accepts delivery of the goods, when the appropriate documents (eg bill of lading, invoice) have been presented, an obligation of great practical importance because the supplier does not want the buyer to refuse the goods after they have been shipped to the buyer's country, the seller already having paid freight charges to get them there

- Arranges and pays for transportation and insurance from the port of destination to their final destination in the buyer's country

5.1.5 CIF: cost, insurance and freight

Cost, insurance and freight is similar to CFR, with the exception that it is the seller, not the buyer, who must arrange and pay for the insurance of the goods to the port of destination.

The obligations of the seller are therefore the same as for CFR with some additional responsibilities.

- Arranges for insurance of the goods from the port of shipment to the port of destination (the amount of insurance cover is often the CIF value of the goods plus 10%)

- Pays the insurance premium

- Provides the buyer with the insurance policy or certificate

The buyer's obligations are the same as for CFR, with the exception that he does not have to pay for the insurance of the goods between the port of shipment and the port of destination.

5.1.6 DDP: delivered duty paid

The seller must pay the costs of delivering the goods to the named destination, having paid import duties on the goods. The seller must therefore pay the import duties or taxes, arrange and pay insurance and provide documents that will enable the buyer to take delivery. The buyer's responsibility is to take delivery of the goods at the named destination. This represents the maximum obligation for the seller.

5.2 INCOTERMS and the marketing mix

Delivery and insurance are real costs to those firms who incur them. The marketer can use these items as ways of satisfying customer needs, thereby making the export quotation more attractive.

For example, let us assume that shipping insurance is cheaper to arrange in France say, than in South Africa. A French firm wishing to export to South Africa might wish to arrange the insurance and include it in the export price - the South African importer will benefit from cheaper insurance of goods in transit.

A firm might have its own fleet of delivery lorries. DDP means that the buyer escapes most of the hassle of transportation for a reasonable price.

A firm might be able to offer a menu of export prices. The supplier might agree to a lower price, but on condition that the goods are delivered EXW, in other words that the buyer pays for delivery. If firms wish to indulge in 'over invoicing' then DDP is a means of maximising the value of the invoice.

6 Methods of payment

In the context of international marketing there are three essential elements in taking pricing decisions:

- Determining the basis of calculating the price (eg cost plus, transfer price, local market price and so forth).

- Agreeing with the seller the basis of shipping, insurance etc.

- Agreeing with the seller the method of payment.

The first two of these have been considered above and this section is devoted to the third. There are any number of variations in the methods of payment that may be agreed between the exporting company and its customer but the following provide a series of 'yardsticks'. They are listed in order of increasing risk for the exporter.

- Payment in advance   Lowest risk
- Letters of credit (documentary credits)
- Payment on shipment
- Documentary collections
- Open account trading   Highest risk

6.1 Payment in advance

The most secure method of payment for an exporter is to obtain payment in full in advance of shipping the goods. He will then not have any risk that the foreign buyer will refuse to pay or be unable to pay for goods that have already been shipped to him. Payment in advance also means that the exporter, by not giving the buyer any credit, does not have to finance the sale himself for a credit period.

Payment in advance gives security to the exporter. The obvious drawback to payment in advance is the risk for the buyer that the exporter will not actually despatch the goods, or if he does despatch them, that they will not arrive in the required condition or to the right specification. It also means that the buyer is

Developing Global Marketing Strategy

financing the sale for some time before he takes physical possession of the goods. As you may imagine, 100% cash with order is not a common form of payment. However, it is quite common for the overseas buyer to pay a cash deposit in advance, and then to pay the balance by another method.

6.2 Letters of credit (documentary credits)

The documentary credit system is a customary method of payment in international trade.

Definition

Documentary credits involve banks giving a guarantee of payment to the exporter provided that the exporter complies with various terms and conditions (such as providing specified documents to a bank for checking after shipment of the goods and shipping the goods within a certain time).

In brief, the documentary credit system works as follows.

Stage 1   The buyer and seller (exporter) agree a sales contract that includes payment by a documentary credit.

Stage 2   The foreign buyer requests his own ('foreign') bank to issue a letter of credit in favour of the buyer.

Stage 3   The foreign bank issues a letter of credit in favour of the buyer. By doing so it is guaranteeing payment to the exporter provided that the latter complies with the terms and conditions.

Stage 4   The foreign bank asks a bank in the exporter's country to advise the credit to the exporter. This does not necessarily involve the latter in making any commitment to the exporter to add its own guarantee of payment but if it does so it is then known as a confirmed letter of credit.

Stage 5   After shipping the goods the exporter has to present the documents specified by the letter of credit. These normally include a commercial invoice, bill of lading and insurance certificate.

Stage 6   Provided the documents are in order the exporter arranges for payment to be made (either immediately or on deferred terms) by the buyer's bank.

These are the two basic types of documentary credit.

- A revocable credit allows the foreign buyer to amend the credit (or even cancel it) without giving prior notice to the exporter. In other words, for the exporter this is high risk.

- An irrevocable credit can be amended or cancelled only with the agreement of all parties to the credit (ie buyer, issuing bank, advising bank and exporter). It therefore gives greater security to the exporter because the issuing bank's guarantee remains in force, even if the buyer changes his mind.

Although the irrevocable letter of credit gives the exporter the guarantee of a foreign bank, the exporter may not be entirely happy relying on a bank about which the exporter has little information in a country that might, for example, impose exchange controls.

This can be overcome by confirmation (or additional guarantee of payment) by a bank in the exporter's own country. In the UK an exporter might ask for a credit to be confirmed by a first class London bank (eg one of the clearers). A confirmed irrevocable letter of credit thus gives greater security to the exporter because the guarantee is given by a bank in the exporter's own country. It makes little difference to the buyer.

Not surprisingly, all this international activity costs money. The cost of issuing a letter of credit is usually borne by the buyer, although the buyer may be able to persuade the exporter to bear some or all of the costs and charges. Who pays what costs depends of the relative bargaining strength of the two parties.

6.3 Payment upon shipment of the goods

The exporter and overseas buyer might agree an arrangement whereby the buyer pays for the goods as soon as they are shipped. The exporter would have to notify the buyer of the shipment, giving full details of the shipment, and then expect the buyer to make an immediate payment. The goods will therefore be in transit or at their point of destination when the payment is received.

Security is provided to the exporter because there are documents of title to the shipped goods, and the exporter can arrange to keep these documents until payment has been received. If the buyer does not make the payment on shipment, or if the funds are not cleared, the exporter will still have title to the goods, because he holds the documents of title. However, the goods would now be in transit and so he would have the problem of deciding what to do with them when they arrive at their destination.

6.4 Documentary collections

Documentary collections involve the exporter asking his bank to help with the arrangements for payment, by handling shipping documents as well as a bill of exchange or a cheque. They provide some security of payment for the exporter because the banks involved in a collection should act in accordance with the internationally accepted rules that apply to collections.

A significant feature of a documentary collection is that if a bank is instructed to handle commercial documents which include a bill of lading, the exporter can keep control over the goods until the foreign buyer has either paid for them or accepted a bill of exchange. This is because the bill of lading is a document of title and a full set of the signed originals of this document can be kept by the bank.

6.5 Open account trading

Open account trading is where payment is received after delivery without any guarantee and is the most risky method of trading for exporters, although it accounts for some 70% of UK exports.

However, when trading relations between an exporter and buyer are well established, and risk is considered low or non existent, then open account trading might be perfectly acceptable without excessive risk for the exporter.

Open account trading has developed quite extensively between countries in the EU for these reasons. A further point to bear in mind is that a vast amount of world trade occurs between fellow members of multinational groups (eg a vehicle manufacturer assembling cars in country A with engines and gearboxes bought from its sister companies in countries B and C).

It is fair to ask why an exporter, who is worried about the risk of non payment, does not insist on payment in advance, payment on shipment or a letter of credit every time he or she exports goods. The answer is, of course, that it takes two parties to trade, the exporter and the importer.

The exporter is often unable to get payment terms which would be desirable in an ideal world, because of the fierce competition in international trade. Another supplier might offer more favourable terms.

6.6 Bills of exchange

This can be coupled with a number of different payment methods but is such an important and established means of payment in international trade that it is worthy of consideration in its own right. It is similar to a cheque. (By writing a cheque, you are asking your bank to pay a sum to the payee whose name is written on the cheque.)

Thus it is basically a request for payment, normally from the supplier (exporter) to the customer. A bill known as a 'term bill' allows the recipient a period of credit before payment. When such a bill is sent to the customer he is expected to sign it as the 'acceptor' and the bill becomes an 'accepted' bill.

7 Countertrade and leasing

A relatively new phenomena in modern global marketing is countertrade. This is included here under the broad heading of price since they are mainly concerned with payment and export financing.

> ### Definition
>
> A UN commission defines **countertrade** as '...commercial transactions in which provisions are made, in one or a series of related contracts, for payment by deliveries of goods and/or services in addition to, or in place of financial settlement'.

Countertrade is a reaction to adverse developments in the world economy:

- Countries with large external debts can acquire imports which are otherwise denied them for financing reasons

- Firms facing excess world capacity and slack demand can gain competitive edge by accepting customer's outputs as payment

- Reciprocal trading can be established between countries endowed with appropriate scarce raw materials goods

- Countertrade deals can overcome financing difficulties arising from high rates of interest and inflation

- Countertrade can be used as a form of payment

- Countertrade is better than 'no trade'

In simple terms, countertrade is the sale of goods from one party to another subject to a sale in the reverse direction. It can be treated as a modern form of barter. Two separate contracts are normally involved; one for the supply of the goods (or the mix of money and goods) delivered by each party, specifying the quantity, dates and method of payment. There is not normally a legal link between the two contracts and the performance of either is legally independent of the other.

Countertrade is better than nothing, but most firms prefer cash.

(a)   Countertrade is highly complex, time consuming and costly.

(b)   The compensation goods are often unattractive and/or difficult to dispose of.

- For firms having an 'in house' use for the offsets, the problems are concerned with low or inconsistent quality, design and poor delivery.

- For firms without an 'in house' use for the offset goods, the problem lies in difficulties in reselling them.

- The goods taken in compensation might become serious competition in Western firms' own markets.

- Many factors lead to pricing problems in countertrade.

There are generally accepted three types of countertrade options:

(1) **Barter**

- A straightforward exchange of goods without any money transfer

(2) **Compensation deal**

- The export of goods in one direction. The 'payment' of the goods is split into two parts:

    (a) Part payment in cash by the exporter

    (b) For the remainder of the payment the original exporter makes an obligation to purchase some of the buyer's goods

(3) **Buy-back agreement**

- The sale of the machinery, equipment or a turnkey plant to the buyer's production is financed in part by the exporter's purchase of some of the resultant output.

Barter and countertrade present their own challenges to the international marketer. Barter is sometimes the only way to get into a new market or penetrate an existing one. Barter requires the commitment of fewer resources and so can be a good way of 'testing the water'.

It is not certain how much room for expansion is offered by barter and countertrade as countries continue to develop, but there are still a large number of countries where economic difficulties mean that barter and countertrade are still seen as attractive arrangements. Exporters from more highly developed countries may find themselves required to take unrelated goods, but they often consider this a price worth paying given the opportunity to crack a new market.

7.1 Advantage and limitations of countertrade

Advantages	Limitations
• New markets can be developed for a country's products, as 'skills' are often imported with the deal	• Lack of flexibility as transactions are dependent on product availability and countertraded products are often of poor quality
• Surplus and poorer quality products can be sold through countertrade whereas they could not be sold for cash	• Products taken in exchange may not fit with the organisation's trading objectives
• Can strengthen political ties through trade agreements	• Negotiations may be difficult as there are no or little guides to market prices
• Can be used to enter high-risk areas	• Countertrade deals are difficult to evaluate in terms of profitability and through countertrade, organisations may inadvertently create new competition
• Can provide extraordinary profits as it allows organisations to circumvent government restrictions	

Global case study

Brazil's state-run oil producer Petrobras has signed a $10 billion credit line with the state-controlled China Development Bank in exchange for future oil supplies. The deal is the first of its kind for the Brazilian oil major as it seeks to secure financing for its $174.4 billion, five-year investment plan to develop and explore new oil fields. Last month the firm announced it would be raising its five-year investment plan by 55% and has plans to invest $28.6 billion in 2009 alone. China, through the China Development Bank has signed a similar $25 billion countertrade deal with Russian oil producers Rosneft and Transneft (*Trade Finance* 19/02/08). Under the agreement Russia will supply 15 million metric tons (110 million barrels) of crude annually under a 20-year long-term contract to China. Proceeds for the deal will also be used to help build a branch of the East Siberia-Pacific Ocean (ESPO) oil pipeline towards China.

7.2 Leasing

A further alternative to outright purchase, particularly where there is a shortage of capital, is the leasing option. Usually the rental fee covers servicing, the cost of spares and so on.

Leasing arrangements are particularly attractive in countries where investment grants and tax incentives are offered for new plant and machinery, in which case the lessor may take advantage of the tax provisions in a way that the lessee cannot, and share some of the savings.

Activity 2

Why do you feel that knowledge of countertrading should be part of a global marketer's pricing toolkit? Is it relevant for the organisation you are investigating and if so why and what may be the benefits of taking such an approach?

Chapter roundup

1 Examine the role of pricing

- In global marketing, pricing decisions are more complex than in domestic marketing, being affected by a number of additional external factors such as fluctuations in exchange rates, and the use of alternative payment methods.

2 Evaluate the factors influencing global pricing decisions

- Global pricing is one of the most difficult aspects of the development of an effective global marketing strategy.

3 Consider global pricing approaches

- The choice is between standardising the product in all markets to reap the advantages of scale economies in manufacture and, on the other hand, adaptation which gives the advantages of flexible response to local market conditions. Adaptation may be unavoidable in some circumstances and therefore a standardised approach is very difficult to achieve.

- Global marketers must take serious consideration to minimising risk through sales being achieved in higher-risk economies and the organisation must put suitable measure in place.

- Pricing must be an integral element of the overall marketing strategy in order for the organisation to take advantage of opportunities and protect itself against potential threats.

4 Identify terms of sale

- There are a number of internationally accepted standard forms of dividing these costs between buyer and seller. Who pays what, and who is responsible for arranging for the transport, has to be confirmed when the sale contract is agreed.

5 Assess methods of payment

- In the context of international marketing the basis for calculating price and agreeing issues such as shipping, insurance and methods of payment should be agreed from the onset.

6 Consider the issues of countertrade and leasing

- Countertrade (eg barter) is the exchange of goods for goods. It is used by countries when access to hard currency is difficult to find. Generally speaking, it is not preferred as a means of payment and export financing.

1   The answer to this activity will depend largely on the market sector in which your organisation operates. Depending on their positioning, targeting and overall corporate objectives they may well adopt different pricing approaches.

2   Countertrading should be considered because it is a reaction to adverse developments in the world economy and provides an option which may allow some organisations be able to get into a new market or penetrate an existing one. The marketer should be aware of the advantages and limitations of adopting such a strategy and should carefully evaluate under which conditions it could be implemented. Its relevance to your organisation will largely depend on the organisation's corporate objectives, organisational structure and product mix.

Cateora, P. and Graham, J., (2009). *International Marketing*. 14th edition. London: McGraw Hill.

Doole, I. and Lowe, R., (2008). *International Marketing Strategy Analysis, Development and Implementation*. 5th edition. London: Thomson Learning.

Hollensen, S., (2007). *Global Marketing*. 4th edition. London: Prentice Hall.

Kotler, P., (1997). Marketing management: Analysis, Planning, Implementation and Control, 9th edition, New Jersey: Pearson.

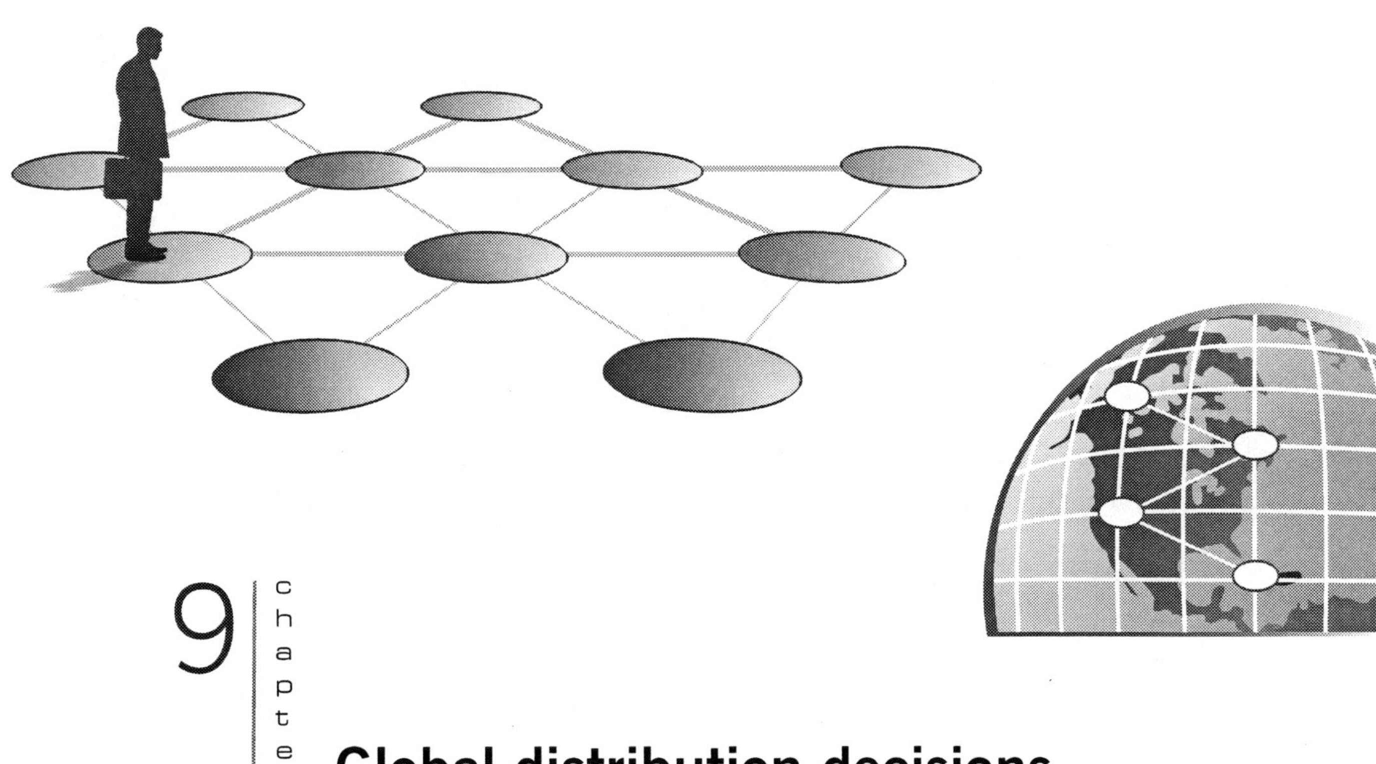

9 chapter

Global distribution decisions

For marketing goals to be achieved a product or service must be made accessible to the target market at an affordable price. This can be a costly process if inadequacies within the distribution structure cannot be overcome. In this session we explore the main issues that must be considered in developing a successful global distribution strategy. We explore the importance of distribution within the total marketing mix, the criteria for selecting, managing and controlling distribution channels and the intermediaries contained within them, as well as an appreciation of the documentation necessary to undertake effective global trade.

We start by setting the scene and the major areas the global marketer must consider in global distribution decision making. We then explore the strategic importance of distribution before considering the options available in channel design and selection and how to manage, select and control distribution channels. Finally we explore global logistics issues and the documents typically used in facilitating global trade.

Contents

Chapter learning outcomes

In this chapter you will cover the following:

- Consider the strategic importance of distribution
- Evaluate channel design and selection decisions
- Identify potential channel structures
- Select, manage and control distribution channels
- Appreciate the importance of global logistics
- Understand documents used in global trade
- Understand the trends and consequences associated with global retailing

Assessment advice

It is now useful to evaluate the distribution strategy for the organisation you are investigating. Consider the options in light of current and past performance and evaluate the distribution strategy you are recommending for the country you are proposing to enter.

1 Setting the scene

For many global marketers, distribution is the less glamorous and 'creative' aspect of the marketing mix. Organisations that use distribution as a means of competitive advantage take distribution very seriously.

As products become more standardised the ability to compete on customer service becomes more vital. It is therefore becoming increasingly important for an organisation to have a well managed and integrated supply chain both within markets and across borders.

As in all aspects of global marketing, overseas distribution increases complexity. The major areas that the global marketer will need to consider are as follows:

Selection of foreign country intermediaries	• Should the organisation use indirect or direct channels? • What type of intermediaries will best serve their needs in the marketplace?
How to build a relationship with intermediaries	• Management and motivation of intermediaries in foreign markets is especially important to organisations trying to build a long term presence
How to deal with varying types of retailing infrastructure across global markets	• An important consideration where retailing is at varying stages of development and there are varying levels of experience and knowledge
How to maximise new and innovative forms of distribution	• Particularly opportunities arising through the Internet and other forms of digital delivery
How to manage the logistics of physically distributing products across foreign markets	• Organisation's need to evaluate the options available and develop a well managed logistics system

2 The strategic importance of distribution

The distribution channel is of strategic importance:

(a)   It is hard to change in the short term, unlike the price or promotion elements in the marketing mix.

     (i)   A distribution channel often involves contractual arrangements with the distributor which cannot be changed easily.

     (ii)   There may be a substantial physical infrastructure involved.

     (iii)   In many respects, distributors are key stakeholders whose needs should be considered, and with whom long-term relationships should ideally be built.

(b)   An organisation's marketing communications will be strongly influenced by the extent to which it is able to obtain wide distribution.

(c)   In overseas markets most marketers will initially be inexperienced. At home, the distribution channel will exist already, whereas in new markets new channels will have to be built up from scratch, in very different conditions. Distribution will often need to be outsourced.

(d)   Distribution is a competitive battlefield and can offer a competitive advantage.

Key issues in distribution are these.

(a)   Coverage and density, in other words the number of sales outlets.

     (i)   Countries like the UK and US allow large stores.

     (ii)   In Japan, there have been restrictions on store size, to protect the livelihoods of small retailers.

(b)   Channel length – the number of intermediaries between producer and consumer.

(c)   Power and alignment. The global marketer has to realise that distribution channel power is not equal in each country. Different roles are played by retailers, wholesalers and agents in each country. For example, wholesalers are most important where retailing is fragmented such as in India for example. In the UK, where, in groceries, concentration of retail power has gone furthest, the major supermarket chains are powerful.

(d)   Logistics and physical distribution.

2.1 Distribution costs

The organisation should remember its basic distribution and logistics strategy is constrained by the distribution cost structure, which may be represented by the following function:

$$D = T + W + I + O + P + S$$

where D = total distribution cost

     T = total transport costs

     W = warehousing costs

     I = inventory costs

     O = order processing and documentation cost

     P = packaging cost

     S = total cost of lost sales for not meeting standards set

In global distribution many of these costs will rise

- **Transport:** longer distances and the use of several modes will raise costs
- **Warehousing:** the organisation might have a warehouse system for each country
- **Stock and inventory** will be higher
- **Packaging** might need to be able to handle long journeys
- **Order processing** and documentation: customs forms, taxes and excise, insurance

The key here is to remember that as some of these cost functions increase, others will decrease. So, for example, larger deliveries to a holding warehouse in your main market may reduce total transport costs through efficiencies or discounts with the freight forwarders or shipping companies, but this may be achieved at the expense of increased warehousing and inventory costs.

2.2 The supply chain

Many multinational enterprises have been getting larger. Some writers are arguing that the trend will continue - so that for many sectors there will be fewer players of world class dominating the field.

There have been, at the same time, much closer links with companies in the supply chain in order to extract best value for money and reduce stockholdings. This has had major consequences on the distribution methods of companies in these supply chains, delivering to their customers on a just in time (JIT) basis.

The change in supply chain linkage is demonstrated in the following model.

- Historically, businesses in the supply chain have operated relatively independently of one another to create value for an ultimate customer. Independence was maintained by buffers of material, capacity and lead-times. This is represented in the 'Traditional' model shown below.

- Market and competitive demands are now, however, compressing lead times and businesses are reducing inventories and excess capacity. Linkages between businesses in the supply chain must therefore become much tighter. This new condition is shown in the 'Integrated supply chain' model.

- Monczka (1996) claims that there seems to be increasing recognition that, in the future, it will be whole supply chains which will compete and not just individual firms. This will continue to have a great impact upon distribution methods.

Supply chain model

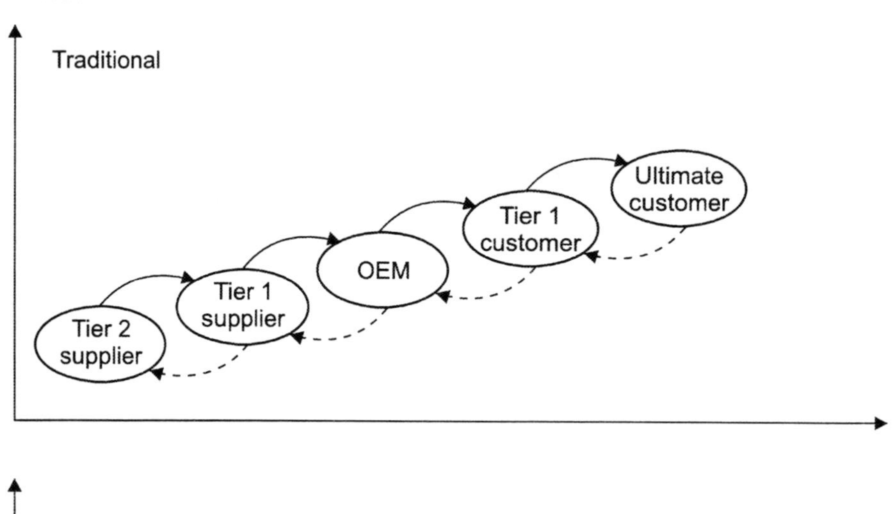

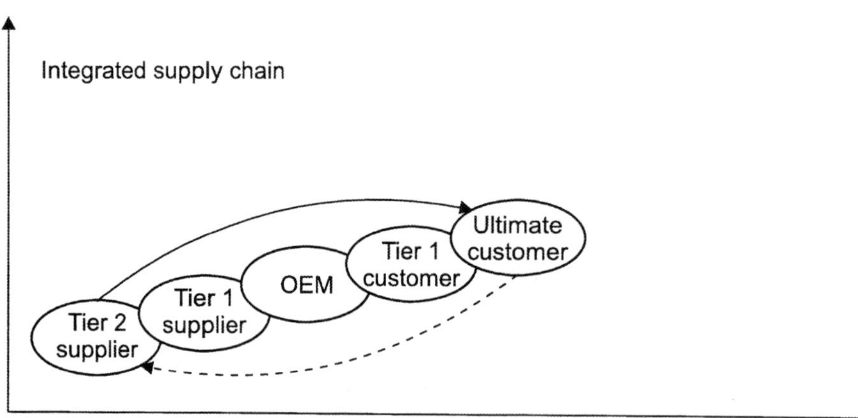

Definition

Supply chain management is about optimising the activities of companies working together to produce goods and services.

Developing Global Marketing Strategy

The aim is to co-ordinate the whole chain, from raw material suppliers to end customers. The chain should be considered as a network rather than a pipeline - a network of vendors support a network of customers, with third parties such as transport firms helping to link the companies.

Global case study

At the centre of Dell's global supplier management programme is a supply chain management system that includes a number of core components:

Certification and Standards

- **ISO 14001 certification.** We require our suppliers to be compliant with ISO 14001, the most widely-recognized standard for environmental management systems, by Jan. 31, 2004 or submit a schedule for achieving certification and obtain Dell approval.

- **OHSAS 18001 certification.** We require our suppliers to be compliant with OHSAS 18001, a prominent European standard for workplace health and safety management systems, by Jan. 31, 2004 or submit a schedule for achieving certification and obtain Dell approval.

Training and Communication

- **Training.** Dell provides training on its supplier expectations during its annual supplier conference as well as updates in quarterly business reviews. In addition, Dell provides training in a variety of relevant areas, such as environmental practices, to suppliers.

- **Supplier social responsibility and environmental responsibility agreements.** As part of every new supplier contracting process, Dell requires suppliers to sign an agreement acknowledging they are aware of and will abide by Dell requirements and principles.

Reviews and Compliance

- **Business reviews.** In order to embed socially responsible behavior into business activities, Dell includes a review of requirements and principles in quarterly business reviews that are held with key suppliers.

- **Self audits.** Dell asks key suppliers to conduct a self-audit to review with Dell management on an annual basis, using a standard scorecard format. These audits are required to be signed by a member of senior management of the supplier.

- **Dell executive oversight and review.** Dell's supply-chain management system is overseen by chief procurement officers and senior vice presidents of Dell Inc. Issues of concern related to supply chain practices are raised in regular operations reviews with these executives and, if appropriate, also raised in Dell's business conduct committee and to the chief executive officer.

- **Board of directors oversight.** Issues of concern related to supply-chain practices will, as appropriate, be raised with the Dell Board of Directors and/or its various committees.

- **Engagement with third parties and non-governmental organizations (NGOs).** Dell will engage with third parties and NGOs as it deems necessary in order to ensure the effective implementation and oversight of its supplier principles.

Correction and Enforcement

- **Correction.** In recognition of the complexity of the world in which we do business, when suppliers fail to meet these principles, Dell and the supplier will create an action plan to ensure future compliance. Performance against the plan to adhere to Dell's standards shall not take more than one year.

- **Enforcement.** Dell reserves the right to terminate at any time, even before corrective plans are developed and implemented, agreements with suppliers that fail to comply with our Supplier Commitment Policy or our Supply Chain Management Requirements.

http://www.dell.com/content/topics/global.aspx/corp/sup_prince/en/supply?c=us&l=en

Managing the supply chain varies from company to company. A company such as Unilever will provide the same product to a range of supermarkets. The way in which the product is delivered, transactions are processed and other parts of the relationship are managed will be different since these competing supermarket chains have their own ways of operating. The focus will need to be on customer interaction, account management, after-sales service and order processing.

Activity 1

The arrival of the global village has had a major impact on the way organisations must view the supply chain. Identify the ways in which your organisation can leverage more value through the supply chain.

3 Channel design and selection decisions

3.1 Types of distribution channels within a market

The participants in a distribution channel have to provide certain services to the customer. A good channel will provide the following.

- Research and information feedback about the market
- Promotion of the goods and services
- Contact and negotiation with prospective buyers
- Storage, sorting, assembling and processing of orders
- Physical distribution
- Financing operations
- Speculative risk taking

The producer's problem is in ensuring that these activities will be carried out. It can be seen that where a weak distributive infrastructure exists within a country, the duties will fall on the exporter and the selected entry organisation, and thus care must be taken in the selection of any partner in global marketing.

3.2 Design and evaluation

The global marketer has to select intermediaries to fulfil all these roles, and to ensure that these duties are carried out within the channel.

Channel design will be affected by the following.

- Product features
- Buyer behaviour and culture
- Competition
- Organisation objectives in the market
- Capital required and costings
- Control, continuity and communication

3.2.1 Product features

The nature of the product being sold will affect the choice of channel members in a particular country.

Perishable goods need short quick turnover channel systems to retain their value. Some goods require pre-sales surveys, advice, installation, after-sales service, maintenance and local spares and accessory availability.

The distribution channel must be able to provide these enhancements efficiently and quickly. In other cases the intermediary will only be required to collect and pass on orders with few other demands or restrictions. Financial services may be required to offer credit in the cases of some high value goods, and the intermediary may be required to provide such a service.

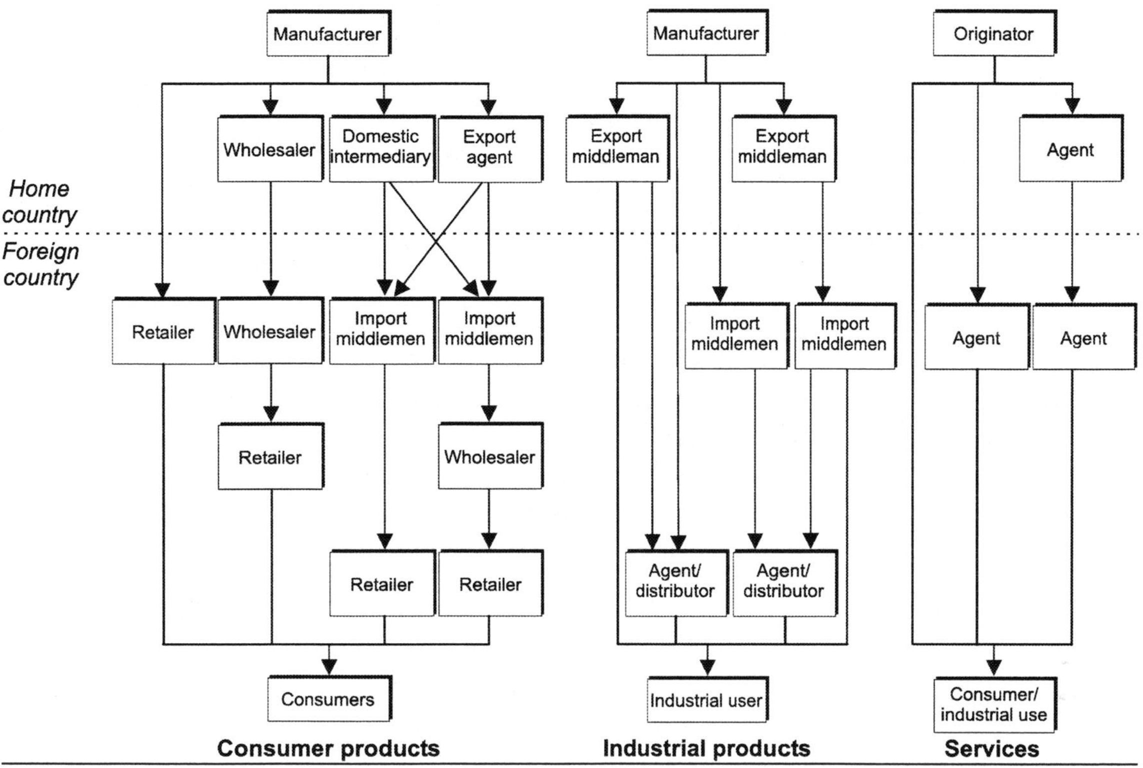

Of course the model above does not automatically consider the role of the Internet and disintermediation meaning that for many products and services it is possible to purchase products as a consumer from foreign organisations directly. This effectively cuts out all other distributors within the processor at the very least blurs/merges their roles.

3.2.2 Buyer behaviour and culture

The social, demographic and economic profile of the target customer varies from country to country.

Each segment exhibits different purchasing habits. A key determinant of this behaviour is the purchasing culture prevalent in the target market. This affects the following.

(a)   The end user preference as to where to buy, frequency of purchase and services required

(b)   The services that the intermediary both expects and is capable of providing, and their purchasing patterns

One example to consider from the onset is the UK retail industry. Through their economic concentration and power, they have generally been able to thrust most of the duties outlined above back onto the manufacturer. In other countries stocking and promotion may be regarded as the legitimate activity of the importing wholesaler/distributor rather than the manufacturer.

In some countries, local legislation may prohibit the ownership of trading organisations by foreign companies (for example, most developing African countries) and the global marketer is either forced into some form of joint ownership or to select a local intermediary to act on their behalf.

Different conditions in different countries will affect the distribution channel. The more intermediaries there are, the higher the price. In Japan for example, there are often several intermediaries who have powerful vested interests.

Preferred distribution methods are often deeply rooted in national culture. For example direct selling is not acceptable in some Asian countries unless it is done via a network of friends and relatives.

Avon Products China says it has now employed 114,000 sales staff in China and is in the process of hiring a further 31,000 recruits just four months after it was awarded its first direct sales license in the country.

Direct sales cosmetics companies have been waiting to tap into what is deemed to be a huge potential market following the lifting of a ban on direct sales in the country at the end of last year.

Currently China is seeing some of the largest industry growth in the world, with almost all cosmetic and toiletry categories reporting sales growth well into double figures, figures that are line with GDP that currently stands at 10.9 per cent.

"We are very pleased that nearly 90 per cent of our beauty boutiques have qualified to act as Service Centres under the government's regulations, indicating that our beauty boutique owners want to be involved in direct selling," said SK Kao, General Manager, Avon China.

Until the new laws were passed to re-introduce direct selling to the China market, Avon had been selling on the market through its beauty boutique retail outlets. Now its direct sales business is being built on those established outlets, a factor that is helping to boost the company's position.

The new model means the company's 5,700 retail outlets offer after-sales services - including order pick-ups, product returns and trials - to sales staffs, as well as providing beauty consultation services to consumers and continuing to sell products at retail.

Avon CEO Andrea Jung is expecting big things to come of the market for the company in the near future. At the beginning of the year she stated that the China market could soon add $1bn to the company's bottom line.

Simon Pitman, 18-Jul-2006

http://www.cosmeticsdesign.com/Products-Markets/Avon-goes-full-throttle-in-China

3.2.3 Competition

Established competition in a market will indicate the usual form of channel.

(a)   Where strong and concentrated competitors exist, this may mean that it will be difficult to attract the better distributors. Typically the competitors may have offered exclusive dealerships in return for an undertaking not to stock competitive goods. Thus a new entrant may find it difficult to attract quality distributors to stock and resell their goods or services.

(b)   The global marketer may be forced to consider new and innovative approaches to trading in a market where the traditional outlets may be blocked. This may entail:

- Entering relationships with other distributors selling to same market
- Online and other forms of direct sales

3.2.4 Organisation objectives in the market

The choice of channel will also depend on what the company wants to achieve in a particular market.

Three types of distribution objective are generally considered: coverage, market share and commitment required from intermediaries.

Coverage

There are three approaches to coverage.

Exclusive distribution	• Involves the selection of high quality intermediaries, who in return for providing local stocks and services are given an exclusive territorial trading right, ensuring that no outlet carrying the same product range will be trading nearby.
	• This approach is used when the global marketer requires most of the functions to be carried out locally to a high standard, through a reputable dealer.
	• It results in a few, strategically located, main dealers in a country giving wide coverage, service and support. It applies to most high value goods requiring good pre- and post-sales support.
Selective distribution	• This strategy involves the use of premium quality dealers selected area by area, and while not given any 'rights', involves the avoidance of too many competitive dealers in an area.
	• Luxury consumer goods, designer clothes, perfumes, high quality durables fall into this category.
Intensive distribution	• In this situation the goods or services involved require little or no pre- or post-sales support and have wide appeal (eg most fast moving consumer goods).
	• Sales can be increased by engaging as many dealers as possible in the distributive network, and the exporter seeks to get as high a market coverage as possible.

Market share

Where an organisation is content with a small share of the market, the use of a local distributor in a region of the market may be appropriate. In some countries trading channels may provide national coverage, but in large countries such as the USA and India, the availability of national coverage through one or two organisations is almost impossible. Region by region distributorships need to be developed to obtain nationwide coverage.

Commitment

This refers to the degree of commitment desired. Where an exporter either has little financial resources, or prefers not to invest that resource in a particular country, the preference will be for independent distributors that have no financial tie with the exporter. On the other hand, a greater degree of control in the channel can be exercised if the exporter has some degree of investment in the distributor network.

3.2.5 Capital required and costings

An organisation needs to assess the relative cost of each channel, the consequences on cash flow and the capital required

Relative costs of each channel	• Generally accepted that it will be cheaper to use agents that establish your own sales force
	• But there will be less control and commitment from agents unless you can find a meaningful way to control and motivate the agent that is mutually beneficial
Consequences on cash flow	• If an organisation uses wholesalers or distributors it takes control of the good and the risks which has a positive impact on cashflow
	• If the organisation wants to deal direct with the retailer or the consumer it will have more capital and resources tied up in managing the distribution channel rather than developing the market
Capital requirement	• Direct distribution systems need more capital injected to establish them both on a recurring and non-recurring basis
	• An organisation needs to evaluate whether it can raise finance, access grants and can repatriate earnings

3.2.6 Control, continuity and communication

If an organisation is seeking to build a global competitive advantage in providing a quality service world-wide then the design and development of channels to achieve this is critical.

It is this requirement by the manufacturer and the pressure put on margins that has led to the development of channel integration. This is the process of incorporating all channel members into one channel system and uniting them under one leadership and set of aligned objectives.

There are two types of integration: vertical integration and horizontal integration.

Activity 2

What are the factors that affect the length, width and number of marketing channels for your organisation?

Definition

Vertical integration is seeking control of members at different levels of the channel.

Horizontal integration is seeking control of channel members at the same level of the channel (ie competitors).

Integration can be achieved either through acquisitions or establishing tight co-operative alliances for mutual competitive advantage.

While integration can be longer to achieve it can result in assured member loyalty and longer-term commitment.

Vertical integration

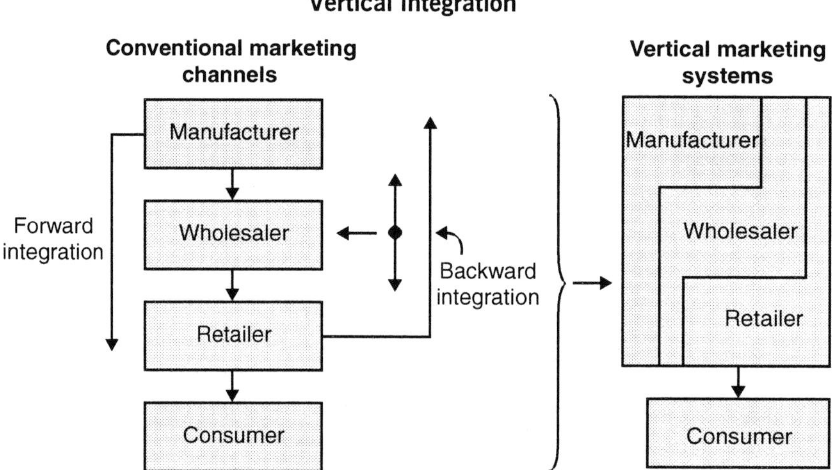

3.3 Mode of entry

As discussed above one way of entering an overseas market is through acquisition, and using the acquisition as a distribution channel.

Channels of distribution for goods which are unprofitable to use should be dealt with and either abandoned in favour of more profitable channels or made profitable by cutting costs or increasing minimum order sizes.

An important aspect of the distribution channel is communication. Good communication is affected by factors such as geographic distance and cultural differences. Religious festivals, for example, in some countries could mean that business cannot happen at all during certain times. Summer holidays in certain countries tend to be taken by a large proportion of the population all at the same time.

Study buddy

Globalisation has had a major impact on organisations' distribution methods. Identify with your Study buddy what you feel between you are the top four factors involved and explain how each has had an impact on distribution. Then together share your thoughts with the wider cohort through the discussion forum.

4 Channel structure

Definition

Channel structure refers to the series of trading intermediaries used to transfer title and physical possession from manufacturer to end user. Sometimes this is referred to as a **trading channel**.

The actual structure of a trading channel is dependent on many factors, which determine the economic viability and efficiency of that form of trading.

A key factor is the ability of a trader to solve the 'discrepancy of assortment'. The discrepancy of assortment refers to the difference between the variety and quantity of goods that are available at the manufacturing level, compared with the requirements at the user level.

Generally, producers wish to produce a very narrow variety of goods and in bulk. End users, on the other hand, prefer to buy a variety of goods in small quantities. A further consideration is the cost of handling and servicing an order compared to the margin available. Thus small orders tend to be unprofitable.

4.1 Consumer trading channels

Consumer trading channels generally require a variety of goods in small quantities, and progressively the trading channel both accumulates a wider variety of goods and breaks down the trade quantities to smaller units.

Thus while an importer may import wine in bulk and then sell it by the case to the retailer, the retailer will sell wine by the bottle and other associated goods that will appeal to the consumer, and provide a sufficiently large order to make the cost of trading worthwhile.

Because the discrepancy is so large, consumer trading channels tend to be longer, with more intermediaries. In some developing countries, where the cost of even a standard pack of a given product may be prohibitive, the local market traders may even break open consumer packs and sell the contents in smaller units.

Servicing rural markets in India involves ensuring availability of products through a sound distribution network, overcoming prevalent attitudes and habits of rural customers and creating brand awareness.

Price-sensitivity is another key issue. Rural income levels are largely dependent on the vagaries of monsoon, and demand is not easy to predict. Most of the companies targeting rural India have started tinkering with pack sizes and creating new price points in order to reach out to rural consumers since a significant portion of the rural population are daily wage workers.

Thus, sachets and miniature packs, as in the case of shampoo sachets priced at Re 1 and Rs 2 or toothpaste at Rs 10, have become the order of the day in hinterland India and help improve market penetration.

Yet, driving consumption of goods in rural areas is not just about lowering prices and increasing volumes but also about product innovation and developing indigenous products to cater to their demands. For example, soap makers use advanced technology to coat one side of the soap bar with plastic to prevent it from wearing out quickly.

In order to efficiently and cost-effectively target the rural markets, companies have to cover many independent retailers since in these areas, the retailer influences purchase decisions and stock a single brand in a product category. In such an environment, being first on the shelf and developing a privileged relationship with the retailer is a source of competitive advantage to consumer good companies.

4.2 Business trading channels

Unlike consumer trading channels, trading on a business-to-business basis usually requires a narrow variety of goods from specialist suppliers in larger quantities. Thus the problems of discrepancy of assortment are less and the trading channels shorter, often involving only one or two intermediaries between the originator and the end user.

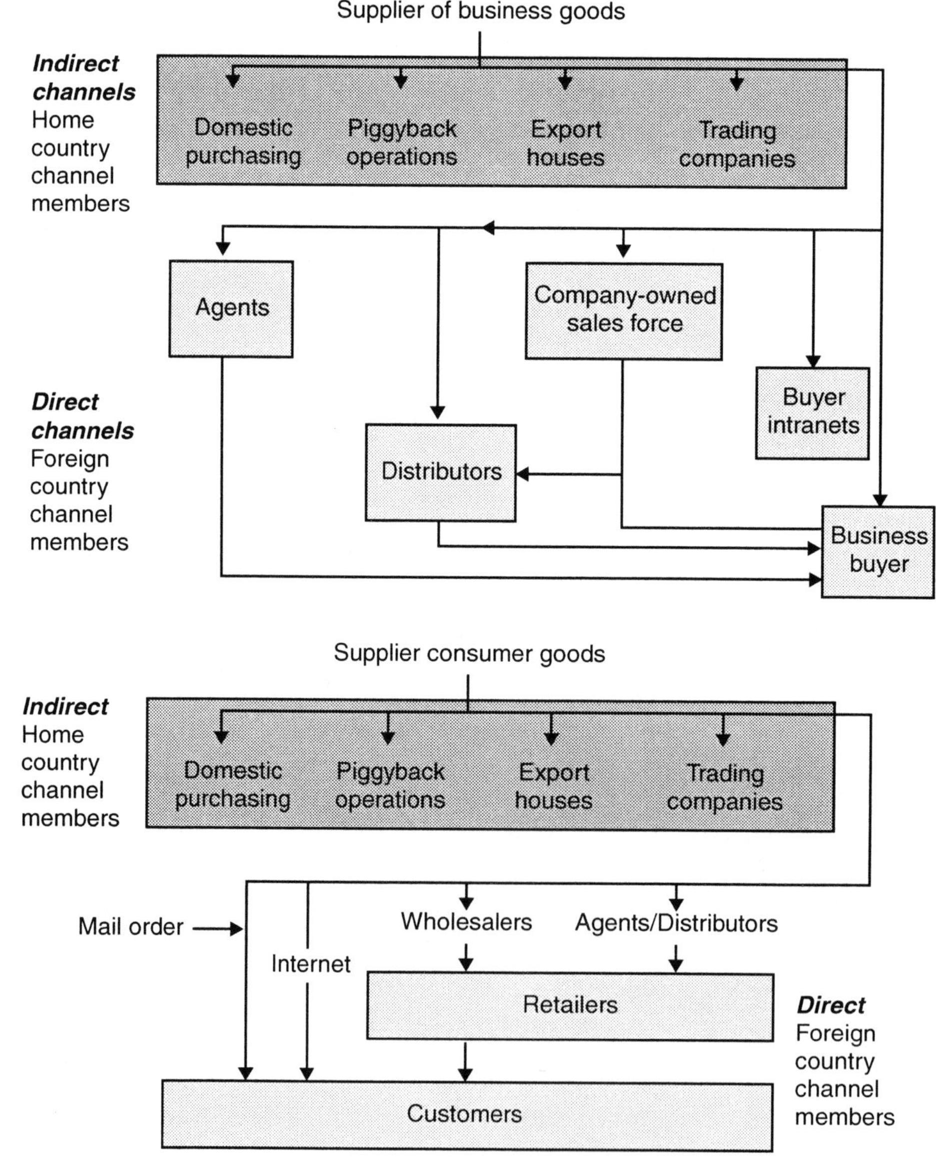

5 Selecting, managing and controlling distribution channels

5.1 Selecting a distribution channel

The following table shows the most important criteria for selecting the distribution channel.

Internal factors	• Cost of ongoing distribution and margins to distributors • Capital investment required • Control over the channel • Coverage of the market • Character of the distribution: changes in the market • Continuity in the relationship
External factors	• Customers • Cultural differences between channel members, expectations of each other • Competitors: how do they distribute? What influence do they have? • Company objectives • Communications

5.2 Selecting agents and distributors

In practice the terms 'agent' and 'distributor' are often used interchangeably. Distributors do take title, they therefore buy and sell goods and are rewarded by way of profits (or losses).

Agents may be either commission agents or stocking agents. Commission agents do not hold stocks. They sell from samples or catalogues. They pass on orders received and the producer delivers directly to the customer. Stocking agents do hold stocks, deliver to the customer but nevertheless are remunerated by commission.

Distributors buy and sell for profit. They hold stocks and are usually given a degree of exclusivity. They often perform after-sales services and other duties on behalf of the producer.

Organisations use overseas agents since they offer the exporter considerable benefits.

- The exporter gains the benefits of the agent's market knowledge and contacts
- It may get rapid results
- The agent usually sells complementary goods
- Investment is minimal
- There is little or no political risk
- The producer gains exporting and market experience

However, the use of agents also involves various disadvantages.

(a) There are likely to be problems in gaining the agent's full commitment since the agent may be selling competing products

(b) The agent may be unwilling to invest the necessary time and effort in developing the exporter's sales if it does not gain immediate results

(c) Many agents are not large enough to serve an entire national market

(d) If the exporter achieves significant market penetration, it may become more economic to open a branch office there. Agents should be used only if they are best suited to achieving the exporter's objectives

Finding satisfactory agents may involve varying degrees of difficulty. The names and addresses of potential agents can be found fairly easily, however, since they are available from many sources including the following.

- Trade associations
- Agents' associations
- Chambers of Commerce
- Banks
- Embassies
- Advertisements in the relevant press
- Word of mouth
- Trade shows, networking events, magazines and online resources

The table below (Hollensen 2007) shows the most important criteria for selecting agents or distributors.

Criteria for evaluating foreign distributors

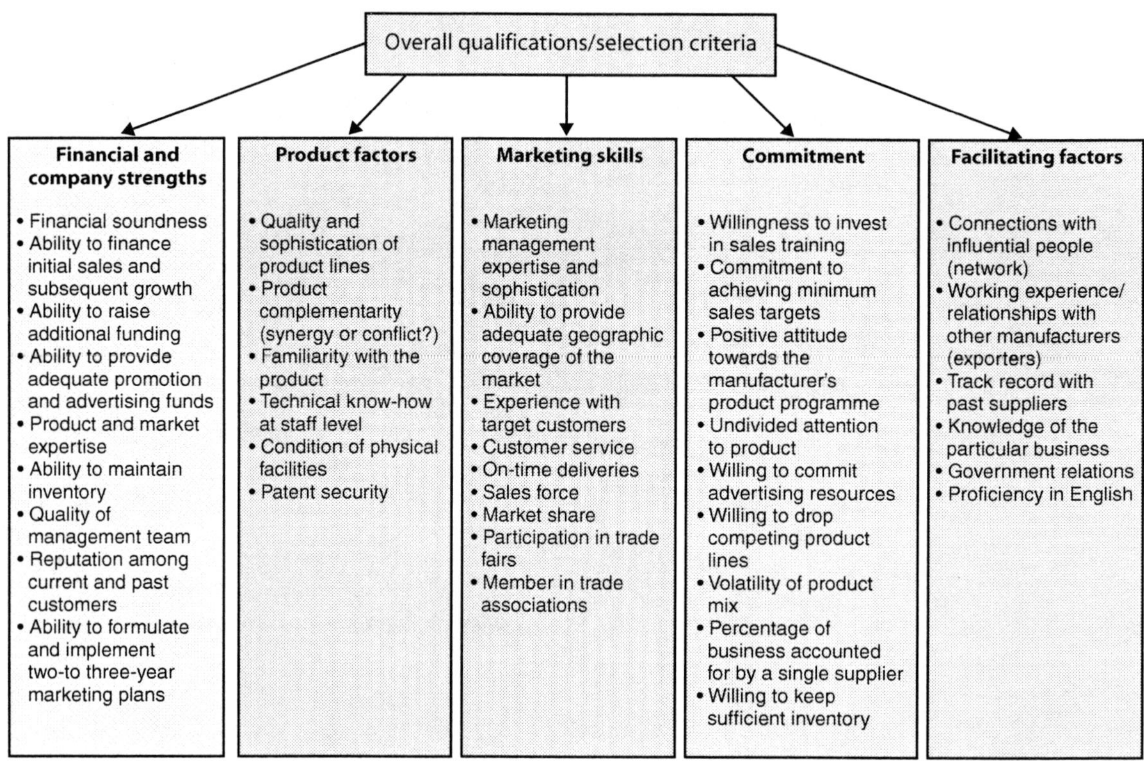

Overall qualifications/selection criteria				
Financial and company strengths	**Product factors**	**Marketing skills**	**Commitment**	**Facilitating factors**
• Financial soundness • Ability to finance initial sales and subsequent growth • Ability to raise additional funding • Ability to provide adequate promotion and advertising funds • Product and market expertise • Ability to maintain inventory • Quality of management team • Reputation among current and past customers • Ability to formulate and implement two-to three-year marketing plans	• Quality and sophistication of product lines • Product complementarity (synergy or conflict?) • Familiarity with the product • Technical know-how at staff level • Condition of physical facilities • Patent security	• Marketing management expertise and sophistication • Ability to provide adequate geographic coverage of the market • Experience with target customers • Customer service • On-time deliveries • Sales force • Market share • Participation in trade fairs • Member in trade associations	• Willingness to invest in sales training • Commitment to achieving minimum sales targets • Positive attitude towards the manufacturer's product programme • Undivided attention to product • Willing to commit advertising resources • Willing to drop competing product lines • Volatility of product mix • Percentage of business accounted for by a single supplier • Willing to keep sufficient inventory	• Connections with influential people (network) • Working experience/relationships with other manufacturers (exporters) • Track record with past suppliers • Knowledge of the particular business • Government relations • Proficiency in English

Adapted from Hollensen (2007)

A visit to the market is an essential part of selecting an overseas agent or distributor. Moreover, it allows the company to have full discussions with agency management, sales staff and service personnel. A visit also enables the exporting company's manager to hold discussions with customers and potential customers so as to gain an impression of how they perceive the potential agent.

5.3 The agreement

The agreement is a commercially binding contract negotiated in great detail by the two parties. Agreements normally include all or most of the following features.

(a)   Identity of the parties

(b)   The purpose of the agreement

(c)   The products that are subject to the agreement and any future changes

(d)   The agent's territory

(e)   Exclusivity, for the agent and for the producer

(f)   Duties of the principal, such as promotional support and training of the agent's staff

(g)   Duties of the agent such as minimum turnover required and after sales service

(h)   Agent's commission, dealing with issues such as the percentage rate, any variation across markets and dates payable

(i)   Duration and dates of contract period

(j)   Provision for termination before expiry of contract (for example, breach of contract and bankruptcy)

(k)   Arbitration provisions for settling major disputes

(l)   Authentic text (that is evidence of which text is authentic if the agreement is written in two different languages)

(m)   Specification of the country whose law governs the contract

The above list should be treated as a skeleton checklist. Most of the points mentioned should be considered in far greater detail before any specific agency agreement is finalised.

5.4 Motivating agents and distributors

A major problem for exporters is gaining the motivation and full commitment of agents and distributors. They usually carry the products of many others and they are tempted to commit most of their resources to the products which provide the quickest returns.

There are various methods for motivating overseas agents, of which the following are the most important:

(a)   Regular and fairly frequent personal contact with the agent. This enables both parties to communicate their needs, problems, philosophies, objectives and mutual responsibilities. Moreover, it allows good personal relationships to be developed among the two sets of personnel. Personal contact can be brought about in several ways.

    (i)   Exporters' visits to the market

    (ii)   Inviting agents to the exporter's headquarters

    (iii)   Agents'/distributors' advisory councils. These can take the form of conferences, holiday breaks and conventions paid for by the producer. They are very successful as motivators and they allow agents to share experiences, complaints and successes

(b)   Assuring agents of long-term business relationships with the exporter.

(c)   Provision of attractive commission and other financial incentives.

(d)   Attractive credit terms.

(e)   Provision of cheap development loans.

(f)   Local advertising, promotions and sales support.

(g)   Training for the agent's personnel in marketing, finance, stock control.

(h)   Exclusivity.

(i)   Effective permanent communications - regular contact by telephone, telex, company newspaper etc, providing up-to-date product and company news.

(j)   Threats to discontinue dealings.

5.5 Controlling agents and distributors

Controlling agents and distributors can be difficult but can be reduced substantially if they are carefully selected.

Realistic expectations of performance are necessary through the common development of written and agreed performance objectives. Contracts should be clear and mutually understood and performance should be analysed on a regular basis through periodic personal meetings.

Performance objectives should also be continually reviewed in the light of changing environmental conditions such as an economic recession or fierce competitive activity. However if only poor performance is being achieved then the contract may need to be reconsidered or terminated.

5.6 Termination

Typical reasons for termination of a channel relationship may be because the global marketer has established a sales subsidiary in a country or following poor performance by the existing intermediary.

Open communication is always recommended to avoid the need for termination. And any termination will have a knock-on effect on the service that is provided to customers.

However if termination is required it is important to be aware of the local laws that may apply. In some countries it may be time-consuming and expensive. A notice of termination will need to be given and for the avoidance of doubt any termination clause must be included in the commercial agreement.

5.7 Developing a wholly-owned global sales force

Organisations with expansion plans may eventually establish their own global sales force. There are advantages and disadvantages of adopting such an approach:

Advantages	Disadvantages
• Provides far greater control over sales and marketing effort	• Relatively large resource commitment
• Facilitates formation of closer manufacturer-to-customer relationships	• High exit costs should the organisation decide to withdraw from a particular market
• Will help to identify and exploit new global marketing opportunities	• Increased exposure to unexpected changes in the political and social environment

An organisation may also choose to use its own sales force for key strategic accounts or work in conjunction with intermediaries and strategic partners to build stronger links with customers.

Activity 3

What are the criteria by which you would select and recruit agents and distributors for your organisation? How would you motivate them to distribute your products?

6 Trends in global retailing

Retail structures differ across countries reflecting history, culture, geography and politics. For example in Italy and France the trend is towards specialist food retailers whereas in the USA the trend is towards large superstores that incorporate a wide range of speciality foods.

A major reason for the lack of growth of large-scale retailing in some countries is due to legislation, with protection in some countries of the local independent retailer.

As legislation around the world relaxes, more European retailers are expanding across the world. However for a retailer to succeed in local markets they must understand the impact of consumer behaviour and the differences in buying habits, tastes, fashions, spending and consumption patterns.

Other limitations include shortages of key resources such as land and labour, unfavourable tax and tariff structures, restrictions on trading hours and foreign ownership and impenetrable established supplier relationships.

6.1 Stages of globalisation

The Uppsala Model identified in Chapter 1 has been applied to depict the typical movement by retailers towards globalisation. Retailers typically move from reluctance to cautious global expansion, typically starting with the closest markets.

The globalisation of retailing has produced different styles of operation.

Type of retailer	Features
Global retailer	Vary their format very little across national boundaries, achieving the greatest economies of scale but showing the least local responsiveness.
Multinational retailer	Operate as autonomous entities within each country
Transnational retailer	Organisation seeks to achieve global efficiency while responding to national opportunities and limitations

Global case study

'Kirana stores to continue operations despite entry of retail MNCs in India"

Despite registering handsome initial success with over 35,000 primary members at its cash-and-carry wholesale store at Amritsar, Bharti Walmart agrees that supply chains in India are very complex and take a long time to evolve. Hence, the traditional kirana stores will co-exist in the country for years.

A company spokesman said organised retail has a very small share in the market today, at a mere 5%. What makes the company hopeful is that given the current growth rates, India has the potential to be one of the largest retail markets in the world within a decade.

A senior company official said India is one of the largest organic growth markets in the world. "India has a growing middle class and a growing consumer attitude. Consumers in India are now beginning to appreciate the modern shopping experience and the company believes it can play an important role towards enhancing this experience," he said.

The company is expected to open 10 to 15 wholesale cash-and-carry facilities and employ approximately 5,000 people in the next three years. Officials say the company's successful experience in Amritsar has shown that there is clearly a demand gap for B2B services.

Over the next three years, Bharti Walmart plans to open 10 to 15 wholesale cash-and-carry facilities. *"We are now in the process of opening more stores in the state as we expand our wholesale cash-and-carry network. Punjab is a great market and is full of unexplored opportunities, besides acting as a hub for various other places,"* he said.

http://www.financialexpress.com/news/kirana-stores-to-continue-operations-despite-entry-of-retail-mncs-in-india/571841/

Accessed 31 March 2010

6.2 Marketing implications for the development of global distribution strategies

The globalisation of retailing has a number of implications of distribution strategies of international organisations:

- Power shifts in supply chains towards retailers
- Intense competition with significant buyer power across boundaries
- New technology facilitating global sourcing and transactions
- Demanding consumers expecting ever higher levels of customer service

The consequence of this is that power is moving from the supplier to the consumer, highlighting the importance of an understanding of the consumer and their behaviour. Consequently there is increasing pressure on suppliers to improve the quality of service.

Retailers are increasingly demanding:

- Streamlined and flexible supply chains
- Guaranteed quality, reliability and delivery across global markets
- Close relationships with intermediaries in the supply chain
- Suppliers who can meet global sourcing requirements

Global case study

Regionalism in China

Multinational corporations (MNCs), typically used to operating at national and multinational levels, face unfamiliar supply chain fragmentation and local protectionism within China. They are confronted with the consequences of centuries of provincial autonomy and self-sufficiency. But modernisation, which brings with it the potential for the application of advanced communication and infrastructure, diminishes the rationale for regional self-reliance. Indeed, such self-reliance hinders interprovincial cooperation, particularly interprovincial trade.

Local administrative bodies and physical infrastructure built to protect local interests pose difficulties for road transportation, private and commercial trucking, Customs clearance for imports, and interprovincial purchasing, whether distribution or wholesale.

Such regionalism also affects domestic regulatory control of retail consumer and agricultural products. For example, health and sanitation certificates with local conditions apply to nationally approved food products, and local laws may require local wholesale purchases of alcohol and tobacco products. Local governments can use both of these means to regulate or hinder the interprovincial transportation and distribution of the products they cover.

Furthermore, regional or local governments may not operate in concert with central-government mandates or laws. Local governments can pass rules and regulations at any time without notice. This situation creates a confusing and costly business environment for companies trying to operate on a national level. Ideally, distribution and supply chains can be managed to achieve optimum efficiency and cost. Today in China, regional fragmentation of the supply chain due to local protectionism decreases efficiency and increases costs—resulting in fewer choices or higher prices for consumers.

http://chinabusinessreview.com/public/0309/wal-mart.html

Accessed 31 March 2010

GMN viewpoint

GMN Module Advisor Svend Hollensen discusses a highly relevant issue associated with the value chain for commodities. It also has an impact on corporate social responsibility and the rise of many 'fair trade' initiatives. During presentations at GMN Global conferences Hollensen uses a slide he titles the banana split model. The photograph of a banana is segmented according to the split of profit along the value chain. For the majority of banana farmers, they more than often only receive a small percentage of profits represented by the stock. You will be able to view this slide within the VLE.

Neil Stevenson (our other Module Advisor) notes that the global marketer should safeguard brand reputation by ensuring distributors embrace good practices in relation to sustainability which are consistent with the brand.

7 Global logistics

The cost of physically transporting goods between countries has been estimated to be as high as 25% of the landed price of an item (more typically 15%), and thus represents a significant part of the cost structure of the item involved. But global logistics provide services to the customer in terms of availability, speed of delivery and convenience, which provide significant added value to the goods themselves.

Logistics involves the following.

- Order processing
- Transportation
- Stock management
- Warehousing
- Customer services

In global marketing, the longer distances involved and the problem of dealing with alternative cultures mean that the time between order and delivery can be considerable when compared to locally sourced supplies, and the cost significantly higher.

A key concept in the management of international logistics is that of customer service. Customer service involves defining adequate levels of performance and is concerned with getting the right goods to the right person, at the right time, at a reasonable cost.

7.1 Transportation

Transport between markets is often the easier part of the problem, with good transport systems and infrastructure (roads, ports, airports, railheads etc) between nations.

Within a particular country however both the transport systems and the infrastructure may vary in quality considerably. As we look to the lesser developed countries, the quality of communications, availability of transport facilities, and the physical infrastructure may pose significant problems.

Modes of transport

The global marketer has not only to take into account the cost of transport between countries, but also the cost within the country.

Temporary storage and handling form a significant part of the overall cost of transport. Thus methods of transport which minimise them tend to be more competitive. The development of container systems for road, rail, sea and air has reduced the amount of labour involved in transferring goods between types of transport. The preference is therefore to use a mode of transport which, wherever possible, provides a 'through delivery' to the customer.

7.1.1 Road transport

Road transport has a major advantage in its ability to deliver directly to the customer's premises. Since other forms of transport (rail, air, water) normally use road transport at the beginning and end of the journey, they require handling on the transfer from one form of transport to another. Per tonne mile road transport is more expensive than rail due to the upper size limit on vehicle loads in most countries, and is generally slower on long journeys than air, even taking handling transfer into account.

7.1.2 Rail transport

Where physical conditions permit, the use of rail transport has significant advantages over its main inland rival, road, in that it can carry much larger quantities over long distances.

Per tonne mile, rail is generally far superior to road in cost over distances exceeding 1,500 km. Over shorter distances the advantages of the load capacity are outweighed by the transfer costs on and off the rail network although this needs to be evaluated in terms of the emergence of high speed trains in certain parts of the world.

7.1.3 Water transport

For bulk transfer of items, water transport has similar advantages to rail, and is widely used where geography allows. Where a suitable network of rivers and canals exists, inland waterways can bring the load nearer to the customer.

Again the problem of transfer and handling may make this approach expensive and generally it needs to be combined with other modes of transport. Increasingly nations are recognising the strategic importance of suitable port infrastructures and are developing such facilities in spite of the heavy investment required.

7.1.4 Air transport

Air transport is probably the most expensive of all four options for short journeys.

The economics of air transport, together with handling costs, make air uncompetitive both in time and cost over short distances. The advantage of air transport lies in speed rather than cost, and this can be most effective over longer distances. The load carrying capacity and fuel costs do not make sense for bulky, low unit value loads, but where high value, small unit size loads are being considered, air transport may be viable.

In particular, where early speedy delivery to a market can give significant price advantages (for example, early flowers and vegetables can command a ten-fold price advantage over main crop in many cases) air transport costs can be justified because of the high unit value of the goods.

Similarly where the customer is willing to pay a premium for speedy delivery, for example a spare part for a manufacturing plant, air transport can be justified. It is of course widely used for high value, low volume, perishable items such as pharmaceutical drugs.

The advantage of air transport, speed, can however be lost where delays at the airport in forwarding to the customer erode the time saving.

7.2 Location of warehouses

If an organisation is seeking to build or acquire a new warehouse, it must seek a general area and then a specific site in that area.

Selecting the area will depend on the market potential. To minimise costs and improve delivery service (and thus increase sales), the warehouse should be sited in the middle of an area with high market potential. The size of the warehouse (ie the amount invested) is also likely to influence the extent to which the market potential is exploited.

The choice of site within an area will depend upon, for example:

- The sites available
- Whether the customer will come to the supplier, or *vice versa*
- Local transport facilities (road, rail, etc)
- Future development in the area
- Whether a lease or a freehold is required
- Its geographical position within the market area

7.3 Logistics management

Logistics management includes physical distribution and materials management. It encompasses the inflow of raw materials and goods together with the outflow of finished products.

Logistics management has developed.

(a)   Customer benefits that can be incorporated into the overall product offering because of efficient logistics management

(b)   The cost savings that can be made when a logistics approach is undertaken

(c)   Trends in industrial purchasing that necessarily mean closer links between buyers and sellers, for example just-in-time purchasing and computerised purchasing.

Logistics managers organise inventories, warehouses, purchasing and packaging to produce an efficient and effective overall system. There are benefits to consumers of products that are produced by companies with good logistic management. There is less likelihood of goods being out of stock, delivery should be efficient and overall service quality should be higher.

Contracted-out distribution services require enormous trust, especially in the food business. Sensitive information must be exchanged between the retailer or manufacturer and the distribution company and responsibility for food safety and hygiene must be shared.

However, contracting out offers great benefits in terms of flexibility and the elimination of the need for costly capital investment: hence the success of specialist global distribution companies such as NFC, Christian Salvesen and Hays.

Activity 4

What factors have influenced the way goods and services are delivered by the organisation you are investigating? Do you think the choices made are all the right ones? If not, how would you change things?

7.4 Just-in-time (JIT)

JIT can be considered: a technique for the organisation of work flows, to allow rapid, high quality, flexible production whilst minimising manufacturing waste and stock levels.

JIT has two main aspects.

1   **Just-in-time production**, a system driven by demand for finished products whereby each component on a production line is produced only when needed for the next stage of a job.

2   **Just-in-time purchasing**, which involves matching the receipt of material closely with usage so that raw material inventory is reduced to near-zero levels.

JIT might not even be appropriate in all circumstances.

- It is not always easy to predict patterns of demand.

- JIT makes the organisation far more vulnerable to disruptions in the supply chain. In February 1997, it was reported that a fire at one of Toyota's suppliers *not only paralysed Toyota's production system, but also hit production at Toyota's other hundreds of suppliers*.

- JIT, originated by Toyota, was designed at a time when all of Toyota's manufacturing was done within a 50 km radius of its headquarters. Wide geographical spread, however, makes this difficult.

Finally, JIT was introduced when customers could enforce it on suppliers. Even in Japan, suppliers to some industries are rebelling against the high administrative and transportation costs. JIT is a feature of markets where customers have high bargaining power. The customer is able to save on inventory costs, but some of these costs are inevitably passed on to suppliers. Suppliers had to build up stocks to ensure flexible delivery.

7.5 Problems in physical distribution

Distribution costs are affected by the following factors:

- Transport costs: higher in overseas markets owing to distance and, in poorer countries, low quality physical infrastructure. A greater variety of transport modes are used.

- Warehousing: more sites might be needed in regional centres.

- Stock levels will be higher, as distances are longer, from the manufacturer.

- Order processing might include customs and VAT documentation.

- Packaging: some adjustment might be needed to reflect climatic, geographical, economic, cultural and distribution channel considerations.

- Failure costs include lost sales through lack of available time. Customers in some markets (eg Germany) are very unforgiving if time and quality specifications are not met.

There are trade-offs between these items:

- Faster delivery (eg by air) might reduce the need to have 'buffer' stocks to satisfy customer orders.

- Stronger and more expensive packaging reduces breakage and theft.

- More warehouse space saves transport costs. Warehouses can be used to tailor the product precisely to the market.

- Investment in IT (eg electronic data interchange) can speed order processing, which might reduce delays, and make ordering easier for the customer, too.

Finally, distribution can be designed as part of the entire customer service package. Elements of customer service include the following:

- Speed, punctuality and flexibility of physical delivery.
- Speed, efficiency and accuracy of procedures for ordering, invoicing and claims.
- Condition of goods when delivered: physical condition and conformance to customer orders.
- Flexibility and convenience of order sizes.
- Personnel: sales visits, after-sales support.

8 Documents used in global trade

There are many documents which are used for one purpose or another in global trade. Paperwork is a crucial (and costly) part of exporting. They can be categorised under the following general headings.

- **Transport documents**. A transport document is a document that indicates loading on board or despatch or taking in charge. Its functions are to provide evidence of a contract of carriage, evidence of receipt of the goods (by the carrier) and, in some cases, they are also documents of title, giving the holder of the documents title to the possession of the goods.

- **Commercial documents**. The most important of these is the invoice. Some types of product also require export licences, such as antiques, works of art and military equipment.

- **Insurance documents**.

- **Official documents** required by government regulations.

- **Financial documents** eg the bill of exchange or the promissory note.

The completion of the relevant forms can be carried out 'in house' by the exporter. The purchase of the necessary expertise is justified where frequent export activity is involved since errors can result in laws being broken, customs regulations being violated or the refusal to honour demands for payment.

Developing Global Marketing Strategy

However organisations exporting infrequently, or who are too small to justify permanent export staff, can call on several forms of assistance.

(a)   The export management company will undertake the total management of export activities, including representation, and ensure that documentation and transport are arranged.

(b)   The freight forwarder arranges transport to the foreign customer and will review and prepare the necessary documentation for a fee. The additional advantage of freight forwarding agents is that they often act a 'transport brokers' consolidating loads for transport to various destinations, and thereby able to obtain transport at very competitive costs.

(c)   The customs broker performs a similar service to the freight forwarder, but on the import side of the transaction, arranging for collection, transport, customs clearance and payment.

The figure below shows a typical export procedure requiring a wide array of documentation.

The export procedure

Albaurn *et al.*, 1994, p.419

According to a survey by DHL, six out of ten Western companies experience difficulties with customs red tape, which hampers trade with Central and Eastern Europe. Customs officials at one border insisted on returning a consignment because the computer printout of the shipment details was on the wrong coloured paper, while another exporter was asked to provide the Latin name for a consignment of potatoes despite the fact that the Mediterranean growers had no idea themselves.

Despite this, there have been some improvements. Forty two per cent of the companies interviewed said they felt the customs officials in these countries at least partially realise what businesses are trying to achieve.

Financial Times

8.1 Transport documents

8.1.1 Bill of lading

> **Definition**
>
> A maritime or marine **bill of lading** is a transport document for goods shipped by sea.

In spite of the considerable growth of container transport, the marine bill of lading is still the most common transport document for exporting to Africa, Asia and the Middle East (particularly when payment by the overseas buyer for the goods is arranged through the banking system).

A bill of lading has three separate functions.

1. It is evidence of a contract of carriage between the shipping company and either the exporter or the foreign buyer, to transport the goods by sea (as detailed in the bill of lading).

2. It is a receipt for the goods taken on board ship, and provides some details about the condition of goods received.

3. A bill of lading, once signed by the exporter, is also a document of title, which means that the holder of the bill of lading has the right to possess the goods. This is very important for the payment arrangements.

Indeed the goods will only be released by the shipping company at the port of destination to someone (the buyer's representative) who presents a signed original of the bill of lading.

8.1.2 Airway bill (also called an air consignment note or air freight note)

An air waybill is a waybill for goods transported by air. Like a sea waybill, it is a contract of carriage and a receipt by the airline for goods received into custody, and it is not a document of title. The airline will hand the goods to the consignee at the airport of destination without the consignee having to present an original copy of the waybill.

8.1.3 Road consignment note or truck receipt

A truck receipt or a road consignment note is a receipt issued by a carrier for goods that are to be transported by road. The note also specifies the name and address of the sender and the consignee, the place of delivery and the place and date of taking in charge by the carrier. The note acts as both a receipt and a delivery note.

8.1.4 Combined transport bill of lading

Increasingly nowadays, with the widespread use of containerised transport, goods are transported from a place of 'taking in charge' to a place of 'delivery' in the same container(s), but on different modes of transport.

For example, a container lorry might take a container of goods by road to a seaport where the unopened container will be loaded onto a ship. It will then be shipped to a port of destination, unloaded, transferred to another lorry, and taken by road, still unopened, to a place of delivery inland. The goods, although carried by two or more modes of transport, are shipped under a single contract of carriage and a carrier or freight forwarder will be responsible for the goods over the entire route. A single bill of lading, known as a combined transport bill of lading will be issued by the freight forwarder to cover the entire route from the place of taking in charge to the place of delivery.

8.2 Commercial documentation

The main commercial documents are invoices. There are five types of invoice.

Pro forma invoice	This is a price quotation by an exporter to a potential overseas buyer. There has been no agreement to buy and sell goods at this stage. A pro forma invoice has several uses.
	• The overseas buyer might need to present a pro forma invoice to the government authorities in his country in order to obtain an import licence or the foreign exchange to pay for the goods.
	• It serves as a price quotation, and might include the terms of sale. If the buyer accepts the quotation, there is a firm contract of sale.
	• It can be used to tender for an export contract.
	• It is used when the exporter requires cash with order
Commercial invoice	This is the demand for payment for goods sold.
	• Description of goods, quantity, unit price and total sales price
	• The terms of delivery (eg CIF Sydney Container Terminal)
	• The terms of payment (eg 60 days after date of invoice)
	The commercial invoice should be capable of easy reconciliation with the transport documentation.
Certified invoice	A commercial invoice which also includes a statement by the exporter about the condition of the goods sent, or their country of origin.
	Some form of statement might be provided at the request of the buyer or for the benefit of customs authorities of the buyer's country.
Consular invoice	A commercial invoice which is prepared on a form printed in the exporter's country by the consulate of the buyer's country. It is then stamped by the consulate. The purpose of a consular invoice is to help the government of the buyer's country control imports into the country (eg to prevent 'dumping' of goods). A consular invoice will be used for exports to any country which requires their use.
Other commercial documents	For example weight, quality and contents (for customs).

8.3 Insurance documents

There are three basic types of insurance document.

A letter of insurance or cover note	• This is issued by an insurance broker to provide notice that steps are being taken to issue an insurance policy or certificate.
A certificate of insurance	• Shows the value and details of the shipment, and the risks covered (in an abbreviated form).
	• It is signed by the exporter and the insurance company.
	• Only a certificate of insurance is required when the insurance policy of the exporting company provides 'open cover' for the whole of its export trade.
	• When an exporter takes out open cover with Lloyd's or an insurance company for the whole of its export trade, a certificate of insurance for each individual shipment will be prepared by the exporter on a certificate provided (and pre signed) by the insurance company. In the UK and many other countries, an insured person cannot take legal action against an insurer where the only evidence of a contract of insurance is a certificate of insurance.
	• The legal proof of a contract is provided by the insurance policy itself
An insurance policy	• The insurance policy gives full details of the risks covered, and it is evidence of a contract of insurance.
	• A policy which provides open cover will not specify the details of any particular shipment.

8.4 Documents required by government departments or agencies

Some documents in international trade are provided to satisfy the legal or procedural requirements of the government of the country of export or import or transit. The following list gives examples and is not comprehensive.

(a)   A certificate of origin might be required by the importing country's authorities as evidence of the country from which the goods originated. This may be necessary, for example, if an exporter wishes to claim a preferential level of import duty.

(b)   A certificate of health, quality or inspection might also be required by the importing country's authorities depending upon the type of product. For example, agricultural and animal products may require a certificate of compliance with health regulations.

(c)   A blacklist certificate provides evidence that the goods did not originate in (and were not transported through) a blacklisted country ie a country with which the importing country has stopped normal trading relations.

Activity 5

Consider the main documents that are required by your organisation in the export of their products.

     Developing Global Marketing Strategy

1 Consider the strategic importance of distribution

- The global marketer has a broad range of alternatives for developing an economical, efficient and high volume global distribution system.

- It is important to understand and manage different cultural differences among different members of the channels in different markets, thus making it difficult to achieve a standardised distribution strategy.

2 Evaluate channel design and selection decisions

- The selected system of distribution is largely determined by the host country channel structure than by what the organisation might ideally like to do.

3 Identify potential channel structures

- The actual structure of a trading channel is dependent on many factors, which determine the economic viability and efficiency of that form of trading.

- Many organisations employ agents who have local market knowledge. The agent's needs have to be taken into account and relationships have to be forged that provide mutual benefit to both the organisation and the agent.

- The strategy for intermediaries, agents, sales team or wholly owned subsidiaries should be carefully crafted.

4 Select, manage and control distribution channels

- Successful distribution management of global distribution channels represents a significant challenge to the global marketer. Success is based on high quality strategic decision making as well as consistent and efficient tactical implementation.

5 Appreciate the importance of global logistics

- The global marketer has to realise that distribution channel power is not equal in each country. Different roles are played by retailers, wholesalers and agents in each country. Logistics management includes physical distribution and materials management. It encompasses the inflow of raw materials and goods together with the outflow of finished products.

- Warehousing, transportation and stock management is highly varied in terms of ease and sophistication between different markets.

6 Understand documents used in global trade

- There are many documents which are used for one purpose or another in global trade. Paperwork is a crucial (and costly) part of exporting. They are categorised as transport, commercial, insurance and official documents.

7 Understand the trends and consequences associated with global retailing

- Retail structures differ across countries reflecting history, culture, geography and politics.

- Lack of growth has been as a result of legislation, resource limitations, existing business relationships and taxation systems.

1   Organisations have to increasingly understand how they can generate more value in the supply chain, either upstream or downstream. This will involve backward and forward integration and can be achieved through partnership, joint ventures as well as through establishing wholly owned operations. For example a diamond mine in Ghana could evaluate how it could generate more revenue through becoming a retailer. The extent of your organisation's ability to generate value will depend upon its vision and objectives as well as its capabilities.

2   This answer will largely depend upon your organisation. Some broad factors to consider include:

- Product features
- Buyer behaviour and culture
- Extent of competition
- Organisation objectives in the market
- Capital required and costings
- Control, continuity and communication requirements

3   This will largely depend on the characteristics of your own organisation but generally you should consider the following:

- Financial and company strengths and capabilities
- Product features and extent to which high degree of knowledge is required
- Marketing skills and the extent to which the agent or distributor has access to the market
- Commitment and the willingness to invest in training, marketing support and so on
- Facilitating factors, including access to Government, opinion formers and so on

There are various methods for motivating overseas agents, of which the following are the most important:

- Regular and fairly frequent personal contact with the agent
- Agents'/distributors' conferences and meetings
- Assuring agents of long-term business relationships with the exporter
- Provision of attractive commission and other financial incentives
- Attractive credit terms
- Provision of cheap development loans
- Local advertising, promotions and sales support
- Training
- Exclusivity
- Effective communications

4   The answer will largely depend upon your organisation. Evaluate the methods being used, the advantages and limitations and the opportunities for improving distribution. Consider how the organisation can make more effective use of the Internet.

5   There are many documents which are used for one purpose or another in global trade. Paperwork is a crucial and costly part of exporting. Undertake a paper trail of your organisation to understand the extent of the documentation that has been used.

Albaum, G. et al., (1994). *International Marketing and Export Management*. 2nd edition. Reading: Addison-Wesley.

Hollensen, S (2007). *Global Marketing*. 4th edition. London: Prentice Hall.

Moncika, R.M., (1996). 'Supplier integration: a new level of supply chain management'. *Purchasing*, vol 120, 0.1, p110 – 113.

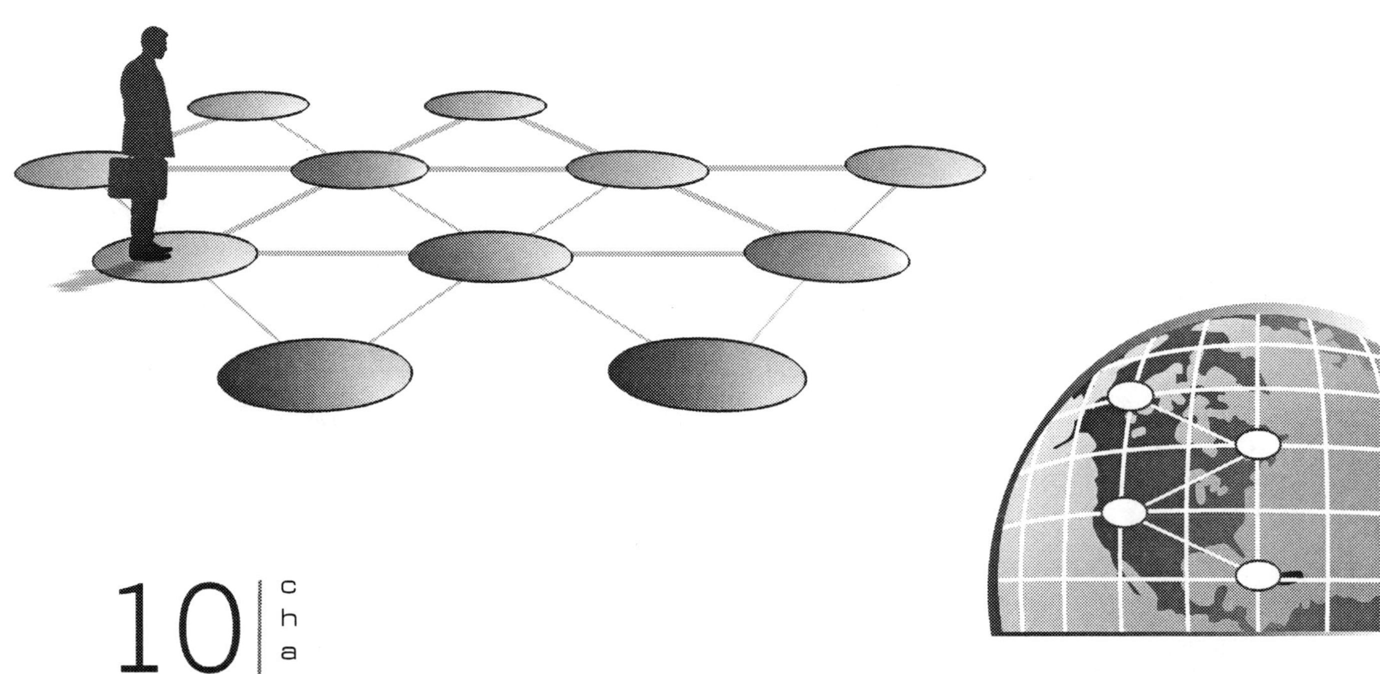

10 chapter

Global marketing management and co-ordination

The overall purpose of this chapter is to summarise the earlier sessions by showing the key steps in developing and controlling the global marketing plan. The main purpose of the global marketing plan is to create sustainable competitive advantage in the global marketplace. While in SMEs this is largely an informal process as an organisation becomes larger its global marketing planning needs to take on a more systematised approach.

First we consider alternative organisational structures that can be considered for the sustainable and effective management of global marketing activities. We evaluate the characteristics of each and the advantages and limitations. We then go on to consider the extent to which activity should be centralised or decentralised with an appreciation of the factors affecting this decision. We consider the options available for controlling the global marketing programme and the measures that can be monitored and evaluated. Finally we consider the role of the global marketing budget and a template than be adopted for ensuring budgetary control.

Contents

Chapter learning outcomes

In this chapter you will cover the following:

- Organisation of global marketing activities
- Control the global marketing programme

> **Assessment advice**
>
> It is now useful to evaluate the distribution strategy for the organisation you are investigating. Consider the options in light of current and past performance and evaluate the distribution strategy you are recommending for the country you are proposing to enter.

1 Setting the scene

Maintaining control of global marketing is of growing concern to organisations, because there is an increasing trend for organisations to be global.

Because plans change with the environment, establishing a control system is vital. With operations in various countries, this is not an easy task. Two aspects of controlling global activities are:

(a)   The control of particular marketing plans in the light of planning and objectives set for them

(b)   The overall control of global marketing strategies to assess how well they are performing and identifying their contribution

Organisational aspects also influence the effectiveness of the global marketing strategy. Whatever form of organisation is chosen, there is still a question as to where decisions are actually taken and different organisations have different approaches to solving this problem, and this will affect planning and control.

These issues are discussed in this session.

2 The organisation of global marketing activities

The way in which global marketing is managed has an impact on the overall ability of the organisation to exploit the opportunities available to it.

The following diagram outlines the evolutionary nature of organisational changes (Hollensen, 2007).

Structural evaluation of international operations

2.1 Functional structure

- Simplest form of structure

- Primarily concerned with efficiency of the organisation

- In an early stage of global development the domestic marketing operation may also have responsibility for global marketing activities

- As global activity intensifies an export or global function may become part of the organisational structure or be a sub-department of sales and marketing

- Particularly suitable for SMEs

Example of the functional structure

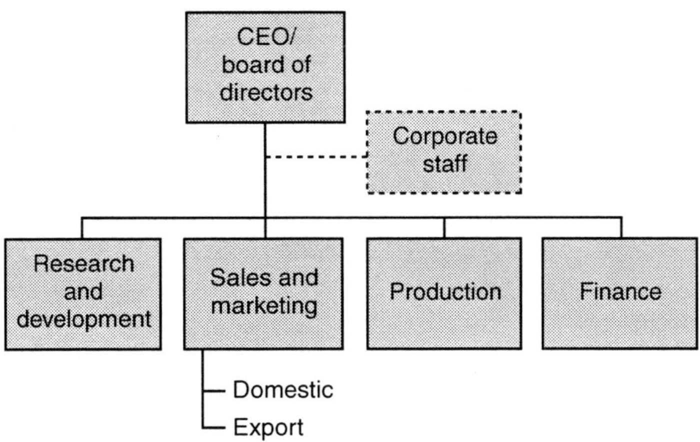

2.2 International divisional structure

- As global sales grow a structure emerges that becomes directly responsible for the development and implementation of the global marketing strategy

- Such an operation would typically involve global experts, research and analysis and authority over global marketing activities

- Manufacturing and other related functions will stay with the domestic operation to take advantage of economies of scale

- Best serves organisations that do not vary significantly in terms of environmental sensitivity and where sales are still much smaller than domestic sales

2.3 Product divisional structure

- More suitable for organisations with more experience in global business and marketing with diversified product lines and extensive R&D activities

- Most appropriate where products have potential for global standardisation

- Cost efficiencies achieved through centralising manufacturing facilities for each product line

- However may suffer from duplication with marketing, R&D and so on in each product division.

- May rely more heavily on centralised (global) marketing leaving little ownership of marketing locally

- May develop independently of each other if there is no communication between them

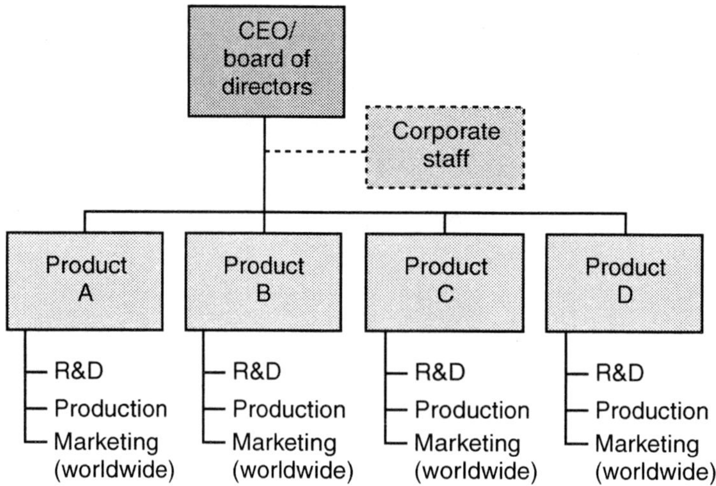

Example of the product structure

2.4 Geographical structure

- Ideal structure if markets vary considerably

- Beneficial where products are largely homogenous and there is a requirement for fast and efficient distribution

- Main advantage is ability to respond quickly and easily to the environmental and market demands of a region or national area with minor adaptations

- Structure encourages adaptive global marketing programmes

- Economies of scale and cost efficiencies can be achieved

- Ensures best use of regional expertise and a degree of area autonomy and ownership

- Duplication may result and the challenge is for the organisation to ensure knowledge transfer and product introductions across regions

- As global sales grow a structure emerges that becomes directly responsible for the development and implementation of the global marketing strategy

- Such an operation would typically involve global experts, research and analysis and authority over global marketing activities

- Manufacturing and other related functions will stay with the domestic operation to take advantage of economies of scale

- Best serves organisations that do not vary significantly in terms of environmental sensitivity and where sales are still much smaller than domestic sales

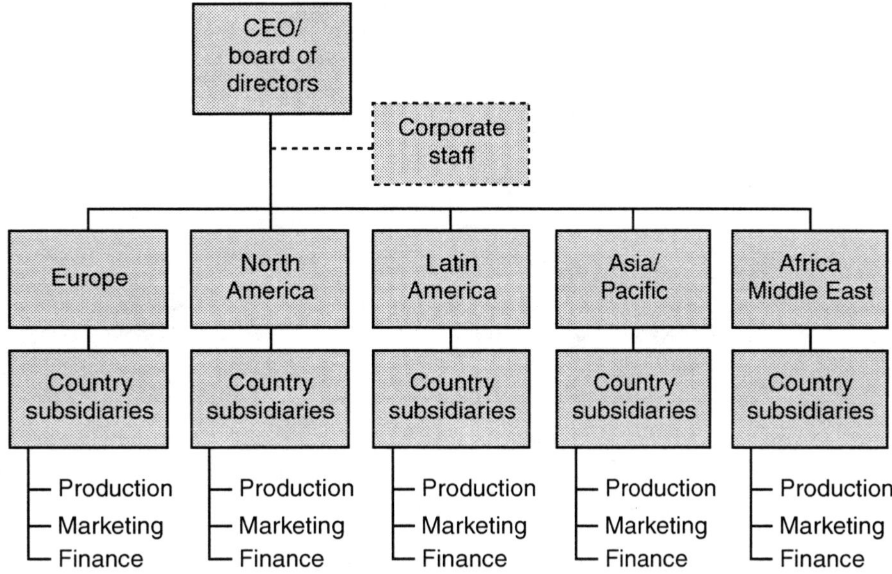

Example of the geographical structure

Within the geographical structure there may be tow further structures:

Regional management centres	• Where sales volume becomes high there will be a need for specialist staff to focus on that region
	• Appropriate where there is a need to create adaptive programmes across the region
Country-based subsidiaries	• Characterised by a high degree of adaptation to local conditions
	• Each subsidiary develop its own activities and is largely autonomous

2.5 Matrix structure

- Consists of two organisational structures (eg geographic and product structures) intersecting each other

- Dual reporting relationships

- Each product division has global responsibilities for its own business and each geographical division has responsibility for the operations in its region or country

- Both division enter into local decision-making and planning processes

- May lead to tensions between divisions between global and local demands and issues

- Dual focus ensures that concerns can be identified and analysed objectively

- Useful for organisations that are both product diversified and geographically spread

- Combining product management and market-oriented approach ensures the organisation can best meet needs of both markets and products

Example of a matrix structure

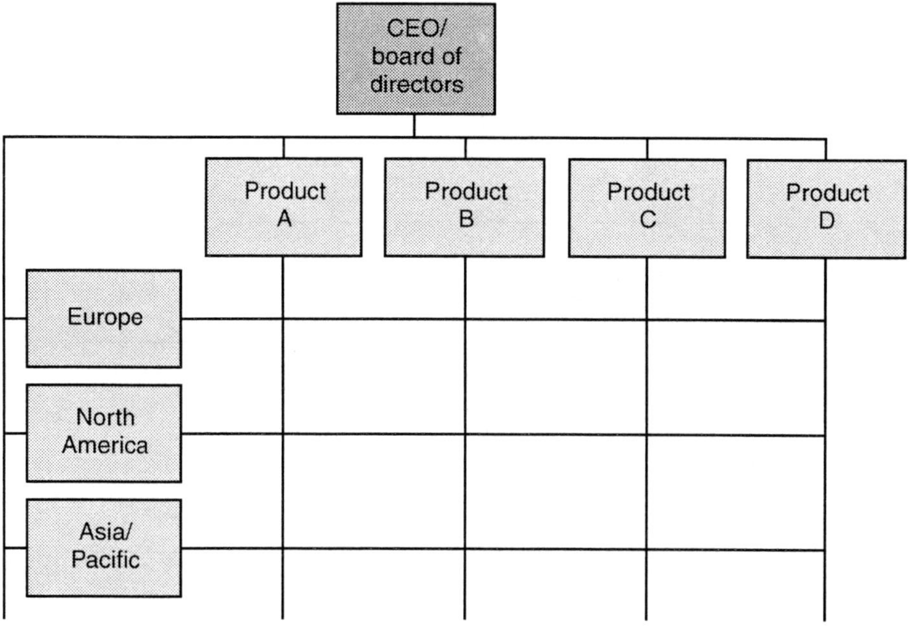

KBR (Kellogg, Brown and Root) is a Texas-based global corporation. It employs over 57,000 people worldwide and is a leading engineering, construction and services company. It supports the energy, hydrocarbon, government services and civil infrastructure sectors. The organisational structure of KBR reflects the market opportunities the business is seeking to maximise all over the world. By dividing up the organisation, into six business units, each can specialise in its own area of expertise, bringing efficiency benefits.

- Upstream – offers engineering, construction, purchasing and related services for energy projects

- Downstream – serves business clients in the petrochemical, refining and coal gasification markets

- KBR Services – provides construction and maintenance services

- KBR Technology – protects the technological property rights of the business. With over 80 years of experience in hi-tech research and development in specialist markets, KBR Technology helps the business to maintain a technological competitive advantage

- KBR Ventures – offers financial investment and management services for companies owning assets of KBR projects

- Government and Infrastructure (G&I) – offers construction, engineering, programme management and services contracting for public and private sector businesses all over the world.

Adapted from: http://www.thetimes100.co.uk/downloads/kbr/kbr_14_full.pdf

2.6 Centralisation and decentralisation

A major concern of management when considering organisational structure is division of labour or specialisation, in other words who does what, where responsibility lies, what authority/control over resources individuals have.

The concept of division of labour has two aspects.

(a)   The span of control (or horizontal division of labour). In international marketing the decision concerning region, product, function, matrix or project structure covers horizontal division of labour.

(b)   The chain or levels of command (ie vertical division of labour). This is the centralisation/decentralisation decision. Here management is concerned with questions of delegation of power and authority and the level at which different types of decisions will be made.

For the global organisation, senior management must consider the division of labour between the headquarters and its foreign market operations.

2.6.1 Factors affecting the extent of decentralisation

Organisation type chosen	The organisation structure types previously discussed vary in the extent to which they lend themselves to a decentralised structure
The nature of the management function	Marketing tends to be a decentralised function, compared with finance for example, because of the importance of understanding local customer needs, responding to environmental differences and changes, developing relationships with customers and so on
The size of subsidiaries	The larger a subsidiary is the more autonomy it will tend to have in its decision making, (ie the more decentralised it will be)

Global case study

Ford banks on China in centralisation strategy

12 February 2009

Ford's decision to shift its marketing team lock, stock and barrel from Thailand to Shanghai is very much a sign of the times. As one of Detroit's stricken 'big three', the car giant is having to think hard about where its priorities are.

First of all, the decision represents another blow to Thailand's already beleaguered marketing communications industry. Last year Unilever decided to shift operations out of Bangkok to Singapore, lured by the promise of "a higher rate of innovation and increased competitiveness" – or bigger tax breaks, in layman's terms. On that occasion, agencies had to shift account teams and reallocate resources away from Thailand. In Ford's case, we can expect to see another exodus of senior talent in the wake of a global client.

Ford uses the same sort of language to rationalise the centralisation of its regional operations as Unilever - "more effective integration and leverage of global assets, and improved management processes that deliver the consistency, efficiency and effectiveness to be competitive".

For a company fighting for survival, every little helps. But from a marketing perspective, the upheaval may still prove a risky proposition. This is no time to lose local relevance. The US auto makers face growing competition in several markets, not least from emerging Asian auto makers such as India's Tata. And in the short term they cannot rely on China, where the middle classes have been hit by the slowdown and car sales are suffering.

For Ford's marketers and agencies, faced with reduced budgets and a move of nearly 2,000 miles, it promises to be a challenging year.

http://www.media.asia/Opinionarticle/2009_02/Leader-Ford-banks-on-China-in-centralisation-strategy/34402

Accessed 26 March 2010

2.6.2 Advantages and disadvantages of centralisation

Advantages of centralisation	• Better planning (providing that 'top down' planning is indeed the most effective)
	• Better co-ordination in achieving overall organisational objectives
	• Optimum use of resources (operational, financial and human)
Disadvantages of centralisation	• Motivation of senior staff at subsidiary level may be adversely affected if they do not have control over decisions or resources which affect their targets
	• Centralisation may lead to misunderstandings and delays in management communications

You work for a small company which sells youth fashion goods products to three different countries of the EU: the UK, the Netherlands, and Germany. There are separate marketing departments and finance departments in each country. A new finance director arrives. The finance director regards the whole arrangement as inefficient, and wishes to return to a centralised structure. What would you consider would be the advantages and disadvantages of such a change in policy?

Activity 1

Discuss the benefits and risks of adopting a matrix organisational structure for your organisation.

2.7 Which organisational structure?

The choice of organisational structure will depend partly on the preferences of the company's top management, and the style they like to adopt, but there are other influences as follows:

- In terms of the marketing mix, the element most likely to be centrally determined is product. Distribution channel decisions may be shared between head office and local management.

- Advertising and promotion decisions may need to be made at a local level for linguistic and cultural reasons.

- When big capital investments are planned, head office should be involved in the decision.

- When cash flow is tight, other strategies must be sacrificed to the paramount concern for short-term survival and attention to cash flow

- Widely diversified organisations, and organisations in industries which are facing rapid change, would be better managed by a greater decentralisation of authority

- Organisations in a single industry which is fairly stable would perhaps be more efficiently managed by a hierarchical, centralised management system.

Activity 2

Consider the organisational structures and evaluate your organisation's current structure. Identify the advantages and limitations of this structure and suggest ways in which the organisation can be improved to enable successful organisation of the global marketing programme.

Developing Global Marketing Strategy

3 The Global Account Management (GAM) organisation

> **Definition**
>
> A **relationship oriented** marketing management approach focusing on dealing with the needs of an important global customer with a global organisation.

The importance of GAM strategies will grow in future because of the consolidation which takes place in most industries. Successful GAM requires an understanding of the logic of both product and service management. It also combines both operational and strategic levels of marketing management.

3.1 Implementation of GAM

Ojasalo (2001) suggests four steps in implementing GAM:

1   Identify the selling organisation's global accounts

It is important to identify which existing or potential accounts are of strategic importance now and in the future

2   Analyse global accounts

This involves making an assessment of the relevant economic and activity aspects of their internal and external environment and relationship history, the account's present and anticipated commitment to the relationship and the alignment of objectives with your organisation at both a strategic and operational level. It will also be useful to assess the organisation's switching costs if for some reason the relationship dissolves.

3   Select suitable strategies for the global account

This will depend on the power positions of the seller and the global account. The selling organisation may not be able to freely select the strategy, may prefer to avoid powerful accounts or may take on less attractive accounts now knowing they will be more powerful in future.

4   Develop operational level capabilities

This may involve for example:

- Joint R&D projects
- Adjustment of the organisational structure through deployment of resources, specialised team and so on
- Recruiting specific personnel with specific skills to manage the account
- Establishing information exchange opportunities
- Offering both organisational and individual level benefits to global accounts

3.2 GAM development

There are typically five stages in the progression of a relationship between the buyer and the seller:

Stage No	Stage Name	Features
1	Pre-GAM	Buying company identified as having key account potential and organisation starts to develop knowledge of the customer's operation and so on
2	Early-GAM	Identifies opportunities for account penetration once account has been won
3	Mid-GAM	Selling organisation has established credibility, contact between the two organisations increases at all levels and assumes greater importance and is perceived as preferred supplier
4	Partnership-GAM	Stage where benefits start to flow. Selling organisation starts to be seen by buying organisation as a strategic external resource. Both organisations engage in joint projects, marketing and share sensitive information freely
5	Synergistic-GAM	Exit barriers have been built up. There will be joint business plans, interfaces at all levels, joint strategies, joint research, reduced costs and so on.

3.3 Organisational set up of GAM

There are three recognised organisational models:

1   Central HQ-HQ negotiation model
2   Balanced negotiation model
3   Decentralised local – local negotiation model

Organisational models	Features
Central HQ-HQ negotiation model	<ul><li>Standardised product</li><li>HQ collects demands from different subsidiaries around the world</li><li>Customer meets with supplier to and the HQ-HQ negotiation takes place</li><li>Discussion come down quickly to a question of the 'right' price</li></ul>
Balanced negotiation model	<ul><li>Central HQ-HQ negotiations are supplemented with local negotiations on a country basis</li><li>HQ-HQ sets the parameters for negotiations on a local basis</li></ul>
Decentralised local – local negotiation model	<ul><li>Negotiations take place on a local basis because the solution requires a high degree of adaptation</li><li>HQs are disconnected from the negotiation process</li><li>While the supplier may be able to achieve higher prices it may also cost more since they will need fulfil different requirements of local subsidiaries</li></ul>

Global case study

Telefonica multinational customers need quick development and deployment of the most suitable fixed and mobile solutions, as well as IT services, anywhere in the world. Across the globe they have more than 1,000 people dedicated to ensuring the customer experience is world class. They deliver a streamlined, complete and consistent service that seamlessly pulls all of the pieces together (in each of their customer's business locations) into one service model built around them.

The Global Account Manager is focused on ensuring that business needs are met at all levels and the Global Account Management team includes an Account, Design and Service Manager.

The Global Account Manager:

- Facilitates engagement and sales interaction at a worldwide level

- Acts as a single point of contact for multinational needs

- Ensures that the technical, design and service requirements are understood and delivered globally

- Gains an intimate understanding of the business to ensure we can support you in achieving your strategic objectives

- Devises and agrees a global account plan

- Project manages multi-country sales solutions

- Standardises and coordinates global information reporting

- Understands customer global technology requirements and sets up informative global 'technology days.'

In tandem with this work, each Local Account Manager will:

- Manage the team in that country to ensure the smooth running of the account
- Devise and propose service development initiatives
- Advises on areas such as service enhancement within their country
- Implements any improvements agreed to on a local level
- Regularly reviews national performance and attends national service review meetings
- Supports the Global Account Manager on implementing the agreed global strategy

http://www.multinationalsolutions.telefonica.com/en/global-account-manager.htm

Accessed 30 March 2010

4 Controlling the global marketing programme

Control is the process of ensuring that the results of operations (including marketing as well as other business functions) conform to established goals. All planning should have a control element along the following general lines.

(a)   Formulation of standards and goals.

(b)   Measurement of actual performance against those standards.

(c)   Procedures for taking corrective action, either by changing the marketing programme, the personnel involved or even adjusting the standards for the future.

Not only is control important to evaluate how we have performed but it also completes the circle of planning by providing the necessary feedback to start the next planning cycle.

The degree of centralisation/decentralisation has important ramifications for the control tasks. For example, a decentralised multinational corporation will not wish to exercise close control, and is likely to find control much harder to implement where it becomes necessary.

4.1 The marketing plan

Both at the Graduate Diploma level and throughout the Developing Integrated Communication Strategy modules you were introduced to the concept of a marketing plan. SOSTAC was described as one such planning model but there are numerous others that you can refer to, the important point is that you should at all times provide a coherent structure to both the process of planning and the plan produced.

Refer back to these earlier modules if you are not clear about the planning process.

4.2 Setting standards and targets

Standards and targets should be clearly defined, understood and accepted by those whose activities are being controlled. This introduces some accountability and 'reins in' some of the more extravagant ideas and so-called 'blue skies' thinking. The question to ask is, "what standards should be adopted?"

Corporate objectives	• Corporate objectives include matters of financial performance.
	   – Profit
	   – Return on capital
	   – Cash flow
	   – Earnings per share
	• Marketing objectives are usually determined to satisfy these corporate concerns and local marketing activities will be designed accordingly
	• Where an organisation is decentralised, each subsidiary may be given its own profit targets, and cash flow targets.

Marketing objectives	• The following are some examples of standards for local marketing activities.
	   – Market research (number and types of studies)
	   – Sales volume (by product line/quarter/year etc)
	   – Market share (by product/quarter/year)
	   – Product (quality control standards)
	   – Distribution (market coverage, dealer support)
	   – Pricing (levels, margins and rigidity or flexibility)
	   – Branch (volume and nature of local advertising/sales promotion)
	   – Selling (local sales force standards)
	   – Customer returns
	   – Number of complaints
	   – 'Share of voice' in international promotions
	   – Consumer awareness

Where practical, headquarters should not impose standards arbitrarily but should use local input, because this will be more up-to-date and will ensure morale and motivation is maintained.

There are many ways of establishing standards.

(a)   Job specifications of marketing personnel give the basic performance requirements.

(b)   The annual planning process will indicate more specific corporate standards.

(c)   Communication (both personal and impersonal) between HQ and subsidiary staff (such as meetings, task forces, internal publications).

4.3 Financial performance

The task of setting financial objectives within a multinational is complex, and several problems must be resolved. The organisation may decide that a single objective should be applied, so that every subsidiary must earn, for example, a specified return on capital employed or profit margin.

On the other hand, because of differing local conditions (such as government policies, taxation, exchange rate fluctuations) it might be more appropriate to apply a different financial target to each subsidiary. For example, the same product may be at different stages in its life cycle in each country.

4.4 International comparisons

If the organisations being compared operate in different countries there will be certain problems for performance measurement.

(a)   Realistic standards need to be set, taking account of local conditions.

(b)   Controllable cash flows. Care must be taken to determine which cash flows are controllable.

(c)   Currency conversion. Care must be exercised to ensure that conversion rates are valid. A firm can lock itself into a particular exchange rate by hedging. Jaguar used this when, as an independent British company, most of its sales were in the US. Hedging instruments were used to protect its profits from any deterioration caused by weakening of its currency receipts.

After developing the global marketing plan its quantification appears in the form of budgets. The budget is the basis for the design of the global marketing control system.

4.5 Obtaining performance information

Monitoring of global marketing performance is difficult since it generally relies more on indirect methods such as reports, as well as direct meetings.

Method of obtaining information	Comment
Reports	• Standardised to allow comparative analysis between subsidiaries • Use an agreed common language and currency • Frequent as necessary to allow proper management • Cover all the information needs of headquarters
Meetings	• Gatherings between HQ executives and subsidiary management allow for more intensive information exchange and monitoring and minimise misunderstandings. • They do however take up time and resources, and are generally not as regular as reports.
Information technology	• Information technology makes it much easier for marketing and financial performance to be monitored closely, given the speed of transmission of e-mail and Internet communications. • Meetings can even be conducted by video-conferencing and Skype – this might prove cheaper than flying and is more environmentally friendly

4.6 Maintaining and enforcing control

The purpose of measuring performance is to enable senior management to take action if the variation between planned and actual performance is unacceptable.

Planning	• If planning is carried out effectively, subsidiary managers will be committed to the goals resulting from the planning process.
Organisation	• The main purpose of organisation as a management function is to facilitate control. Other control methods include the following. • The budget • Corporate culture • Procedures and rules to ensure a standard product or service • Direct intervention in cases of underperformance
Control of intermediaries	• The problem with controlling 'outsiders' is that there is no control by ownership. In the final analysis negative controls such as legal pressures or threats to discontinue relationships can be used, resulting perhaps in loss of business. • The best control is through good selection and by making it clear to intermediaries that their interests and the organisation's coincide.

4.7 Key areas for control in marketing

There are two key types of marketing control that should be considered:

Annual plan control	• Purpose is to determine the extent to which global marketing efforts over the year have been successful.
	• Will centre on measuring and evaluating sales in relation to standard, objectives, target, market share, expenditure and so on.
	• Sales performance is a key element in the annual global marketing plan control. The hierarchy of sales and control is shown in the diagram below.
	• Clearly any variances in achieving sales targets at the organisational level are the result of variances in performance at local level.
	• Variances must be closely monitored and causes must be identified and determined.
Profit control	• Global marketers must be concerned to control their profits and therefore due consideration must be given to the development of the global marketing budget.

Types of marketing control

Types of control	Prime responsibility	Purpose of control	Examples of techniques/approaches
Strategic control	Top management Middle management	To examine if planned results are being achieved	Marketing effectiveness ratings Marketing audit
Efficiency control	Line and staff management Marketing controller	To examine ways of improving the efficiency of marketing	Sales force efficiency Advertising efficiency Distribution efficiency
Annual plan control	Top management Middle management	To examine if planned results are being achieved	Sales analysis Market share analysis Marketing expenses to sales ratio Customer tracking
Profit control (budget control)	Marketing controller	To examine where the company is making and losing money	Profitability by eg product, customer group of trade channel

Adapted from *Marketing Management Analysis, Planning, Implementation and Control*, 9th edition, by Kotler, P. (1997).

Activity 3

Identify the key mechanisms that your organisation uses to obtain information about its marketing performance and ways in which this could be improved.

Developing Global Marketing Strategy

4.8 The global marketing budget

The global marketing budget is the projection of actions and expected results which should be capable of accurate monitoring. It is also an organisational process that involves making forecasts based on the proposed global marketing strategy and programme.

Hollensen (2007) has formulated a table that demonstrates the traditional marketing budget (per country or customer group) and its underlying determinants.

Marketing budget 20XX and its underlying determinants

The most important measures of marketing profitability are:

$$\text{Contribution Margin in \%} \quad \frac{\text{Total Contribution}}{\text{Total Revenue}} \times 100$$

$$\text{Marketing Contribution Margin \%} \quad \frac{\text{Total Marketing Contribution}}{\text{Total Revenue}} \times 100$$

$$\text{Profit Margin \%} \quad \frac{\text{Net Profit (before taxes)}}{\text{Total Revenue}} \times 100$$

If we also have information about the size of assets we can also define the Return on Assets, similar in nature to the Return on Investment

$$\text{Return on Assets (ROA)} \quad \frac{\text{Net Profit (before taxes)}}{\text{Assets}}$$

The table below represents an example global marketing budget for a manufacturer of consumer goods and is used for the following purposes:

An example of an international marketing budget for a manufacturer exporting consumer goods

Internation marketing budget	Europe						America		Asia/Pacific					
	UK		Germany		France		USA		Japan		Korea		Other Markets	
Year = _____	B	A	B	A	B	A	B	A	B	A	B	A	B	A
Net sales (gross sales less trade discounts, allowances, etc.)														
+ Variable costs														
= Contribution 1														
+ Marketing costs:														
Sales costs (salaries, commissions for agents, incentives, travelling, training, conferences)														
Consumer marketing costs (TV commercials, radio, print, sales promotion)														
Trade marketing costs (fairs, exhibitions, in-store promotions, contributions for retailer compaigns)														
= Total contribution 2 (marketing contribution)														

B = budget figures; A = actual

Note: On a short-term (one-year) basis the export managers or country managers are responsible for maximising the actual figures for each country and minimising their deviation from budget figures. The international marketing/director is responsible for maximising the actual figure for the total world and minimising its deviation from the budget figure. Cooperation is required between the country managers and the international marketing manager/director to coordinate and allocate the total marketing resources in an optimum way. Sometimes certain inventory costs and product development costs may also be induced in the total marketing budget.

1 Organisation of global marketing activities

- Allocation of marketing resources among countries and markets to maximise profit. It is the responsibility of the global marketer to maximise total contribution globally.

- Evaluation of country/market performance. It is the responsibility of the country manager to maximise his or her contribution.

- Implementation of a global marketing programme requires an appropriate organisational structure.

- As the scope of the organisation's global marketing strategy increases an organisation's structure will need to be adapted.

- Organisations can choose between treating the international marketing function separately from the domestic marketing function or integrating international marketing with domestic marketing.

2 Control the global marketing programme

- Control is the process of ensuring that the results of operations conform to established goals.

- The global marketing control system will need to overcome international and intercultural differences, and comparing the performance of different countries can be difficult.

- Global marketing performance is generally controlled via reports and meetings. Information technology has an increasingly important role to play.

- The global marketing control system consists of deciding on marketing objectives, setting standards and target, evaluating performance against standards and taking corrective or supportive action.

- The most obvious areas of control relate to the control of the annual global marketing plan and the control of profitability.

- The purpose of the global marketing budget is mainly to allocate marketing resources across countries to maximise worldwide total marketing contribution.

1 The benefits and risks of adopting a matrix organisational structure for your organisation depend upon its existing organisational objectives, structure and culture. The benefits are that each product division has global responsibilities for its own business and each geographical division has responsibility for the operations in its region or country. This dual focus ensures that concerns can be identified and analysed objectively and ensures the organisation can best meet the needs of both markets and products. However it may lead to tensions between divisions between global and local demands and issues.

2 There are advantages and limitations of different organisational structures. Ultimately a structure exists to ensure that the organisation's customers can be served in the best way possible. The answer to this question will largely depend upon the current organisational structure and how well the organisation is meeting its corporate objectives.

3 The monitoring of global marketing performance is difficult since it generally relies more on indirect methods such as reports, as well as direct meetings. There are a number of mechanisms that can be used utilising management reports, meetings and information technology. Some of the key factors are ensuring that management reporting can be efficiently collated and utilised, is consistently reported across the organisation and provides actionable data focused on measuring key performance indicators.

Hollensen, S., (2007). *Global Marketing*. 4th edition. London: Prentice Hall.

Kotler, P., (1997). Marketing management: Analysis, Planning, Implementation and Control, 9th edition, New Jersey: Pearson.

Ojasalo, J., (2001). "Key account management at company and individual levels in business-to-business relationships", *Journal of Business & Industrial Marketing*, Vol. 16 No. 3, pp.199–218.

Index